Crossroads

Crossroads

Environmental Priorities for the Future

Edited by Peter Borrelli

Washington, DC   Covelo, CA

ABOUT ISLAND PRESS

Island Press, a nonprofit organization, publishes, markets, and distributes the most advanced thinking on the conservation of our natural resources—books about soil, land, water, forests, wildlife, and hazardous and toxic wastes. These books are practical tools used by public officials, business and industry leaders, natural resource managers, and concerned citizens working to solve both local and global resource problems.

Founded in 1978, Island Press reorganized in 1984 to meet the increasing demand for substantive books on all resource-related issues. Island Press publishes and distributes under its own imprint and offers these services to other nonprofit organizations.

Funding to support Island Press is provided by The Mary Reynolds Babcock Foundation, The Ford Foundation, The George Gund Foundation, The William and Flora Hewlett Foundation, The Joyce Foundation, The J. M. Kaplan Fund, The John D. and Catherine T. MacArthur Foundation, The Andrew W. Mellon Foundation, Northwest Area Foundation, The Jessie Smith Noyes Foundation, The J. N. Pew, Jr. Charitable Trust, The Rockefeller Brothers Fund, and The Tides Foundation.

For additional information about Island Press publishing services and a catalog of current and forthcoming titles, contact Island Press, P.O. Box 7, Covelo, California 95428.

Cover design: studio grafik.

LIBRARY OF CONGRESS
Library of Congress Cataloging-in-Publication Data

Crossroads: environmental priorities for the future/edited by Peter Borrelli.

Includes index.

ISBN 0-933280-68-8   ISBN 0-933280-67-X (pbk.)

1. Nature conservation. 2. Environmental protection. I. Island Press.

QH75.C74   1988   88-22338
363.7—dc19   CIP

Manufactured in the United States of America
10 9 8 7 6 5 4 3 2

Contents

Introduction

LIKE MANY AMERICANS who at one time or another have experienced the immensity and wonder of nature, I did not have to discover the environmental crisis or be persuaded that it was real. It became self-evident, as the places of my boyhood were transformed into something less wonderful, less peaceful, less clean.

As a boy I had the good fortune of living for a while in Provincetown, Massachusetts, on the edge of the large sand dunes that face the Atlantic Ocean. This was the first landfall of the Pilgrims who explored the area in mid-November 1620, before settling the Plymouth Colony across Cape Cod Bay. Without knowing it, I spent a great deal of time retracing their steps, wandering over the crests and swales of what seemed an endless expanse.

Except for a small but incredibly dense beech forest situated like an oasis about midway between the harbor and the ocean, the land was mostly sand, beach grass, and pitch pine. To the uninitiated it all looked the same. To get your bearings, you had to climb high enough to locate the towering granite memorial to the Pilgrims in town or one of several coastal lighthouses. Even with such sightings, however, it was easy to become disoriented, as the Cape hooks sharply in on itself at the end. (If you have ever tried reading a compass while riding in a car you know the sensation.) The best way to learn your way around (especially as scrambling to the top of a high dune was exhausting) was to become intimately familiar with the shapes and patterns of scattered plant life, fallen logs, and lightning scars, not to mention the position of the sun and direction of the wind. For me it was nature primeval. The thought of it ever changing never crossed my mind.

What I did not know at the time was the extent to which the land had already changed and what the future held. Though I thought of myself as a modern-day Pilgrim lost in an ocean of sand, what the Pilgrims saw was a forest of luxuriant woods: ". . . juniper, birch, holly, vines, some ash, and walnut; wood for the most part open and without underwood." Instead of sand, there was a "crust of the earth a spit's [spade's] depth excellent black earth."

The forest provided tar, turpentine, and potash; wood was used for homes, ship repairs, and fuel. Land was cleared for farming, and cattle and sheep were allowed to graze without restriction. Without tree cover, the fierce winds of the Atlantic began tearing away the soil. By 1714, sand movement was so pronounced that Provincetown and its harbor were nearly obliterated. And so it continued well into the next century until the Commonwealth began efforts to plant beach grass and trees to stabilize the land.

Following World War II, when I was roaming the sands, there was still the chance that nature, given a helping hand, would recover from the first waves of settlement and change, if for no other reason than there was so little of commercial value left to exploit. But Cape Cod was soon to experience a new wave of settlement and exploitation in the form of tourism. By the mid-1950s, what little undeveloped land remained was about to be carved up and paved over when then-Senator John F. Kennedy introduced legislation to create the Cape Cod National Seashore, a forty-mile section of the Cape ending at the point of the beginning—the Provincelands where the Pilgrims first set foot.

During the last fifty years, the story of the Provincelands has been repeated from coast to coast. First came the interstate highways, which sliced and diced the country into little bits. Then came sprawl, followed by urban squalor and decay, as the lifeblood of the cities was drawn to the suburbs. Soon, even the remotest farms and forests of this enormous landscape could feel the pressure of expansion. Golden arches glowed from coast to coast. Save for a great though inadequate park system, which symbolizes our national proclivity to plunder or preserve, the franchising of America was complete.

You do not have to be an ornithologist, oceanographer, or biochemist to understand that the world around us is being abused. To a greater or lesser extent, the selfishness and shortsightedness of our species has always made it so, as George Perkins Marsh documented so well in his 1864 conservation classic *Man and Nature*. Noting that the decline of the Roman Empire was due in part to deforestation, erosion, and land abandonment, he wrote, "There are parts of Asia Minor, of Northern Africa, of Greece, and even of Alpine Europe where the operation of causes set in action by man has brought the face of the earth to a desolation almost as complete as that of the moon."

Marsh went on to observe that in the ensuing centuries man's ability to modify and destroy nature had intensified. "The earth is fast becoming an unfit home for its noblest inhabitants," he wrote, "and another era of equal human crime and human improvidence . . . would reduce it to such a condition of impoverished productiveness, of shattered surface, of climatic excess, as to threaten the deprivation, barbarism, and perhaps even extinction of the species." However, Marsh also believed that *homo sapiens* is a rational species capable of reforming its destructive tendencies. This belief has been the basis of all conservation efforts and the hope of the future.

Sometime during the early sixties, while I was still in college, it struck me that there were only two choices: passive acceptance of the state of diminishment around me (which amounted to no choice at all) or some form of resistance. Fortunately, there were others who felt much the same sense of loss, and visionaries like Rachel Carson, Jacques Cousteau, and David Brower to inspire us to act. The mood of the period was captured eloquently by poet Nancy Newhall and photographer Ansel Adams in their 1960 photo-essay, *This Is the American Earth*. Beneath Adams's now famous black-and-white of the Sierra Nevada taken from Lone Pine, California, Newhall wrote: "This, as citizens, we all inherit. This is ours, to love and live upon, and use wisely down all the generations of the future."

Citizenship, reverence, stewardship, justice. Around these principles environmentalists of the 1960s and '70s rallied. For many reasons which shall become apparent in these pages, it is time to take stock of how faithful we have been to the cause.

This book is divided into two parts. The first deals with current trends within the environmental movement and is comprised of a series of reports that first appeared in *The Amicus Journal* which is published by the Natural Resources Defense Council (NRDC) and of which I am editor. The second is a group of essays by environmental leaders and activists offering their perspectives on the state of the environment and the future of the movement.

John Adams, NRDC's first and only executive director, and John Robinson, a long-time NRDC trustee, came up with the idea for the *Amicus* series by doing what is often the most difficult thing for any individual or institution—looking beyond oneself and asking what is going on. They sensed that the geological plates of the environmental movement were shifting. But why and in what direction?

The series, reported by Tom Turner, staff writer of the Sierra Club Legal Defense Fund and former editor of *Not Man Apart*; Dick Russell, a Boston freelancer and conservationist; and myself, involved interviews with some 200 environmental activists and leaders. Our initial finding was that what some had cited as evidence of a polarization and breakup of the movement actually were symptoms of a healthy, growing, maturing, and diversifying movement. To be sure, the Reagan era has taken its toll. The constant assault on environmental programs from the White House, the Office of Management and Budget, the departments of Interior, Energy, and Agriculture, and the Environmental Protection Agency has consumed much of the energy and resources of the national organizations. But, as with political revolution, the more the administration ignored or resisted genuine concerns, the harder the opposition pressed. The more the president and his advisers insisted it was not acid raining, the harder it rained and the wetter they got.

At the same time, the tectonic shifts Adams and Robinson sensed were occurring were changing the focus and even the composition of the movement. Whereas environmentalists in and out of Congress mostly succeeded in frustrating the administration's efforts to break up the environmental bureaucracy and to privatize public resources, by the mid-'80s the focus of the movement was no longer on Washington. Environmentalists had begun to think globally and act locally in response to such macro issues as global warming and ozone depletion and such micro issues as contamination of drinking water and solid waste management. Though both global and local environmental problems involve national policies and programs, we are entering an age in which shaping international accord is as important as promoting grassroots action.

We also found that many people are becoming disenchanted and discouraged by the bureaucratization of environmental reforms and of the movement itself. (The complaint, though heartfelt, has a melancholy sound, as if the good ol' days of no environmental laws, policies, or agencies are worth remembering.) In large measure, the feeling stems from the inordinate complexity of many of today's issues, especially those dealing with public health. As for the glacial progress of environmental reform, there is no doubt that it has led to burnout, disappointment, and embitterment among citizen activists. A more positive and encouraging finding is that an increasing number of people in this country, and in developing and undeveloped nations, understand the interrelationships of environment, national security, and economics. For the first time, quality of life, the much abused expression of the 1970s, is no longer an abstraction but a social goal. Its proponents, who may be found in the Green movement of Europe, the toxics movement in the United States, and the conservation movements of various African and Latin American nations, are demanding a direct say in the defense of their environment.

Even before the second installment of the series was published, it was apparent that the environmental movement was anything but stagnant and that many leaders and activists had something to say. Among them was Queens College microbiologist Barry Commoner, whose 1971 book, *The Closing Circle,* had a profound influence on the modern environmental movement. The thesis of the book was that industrialized societies have replaced natural and biodegradable substances such as wood, cotton, and manure with synthetic fibers, plastics, and nitrogen fertilizers, which place heavy strains on ecosystems. "We have broken out of the circle of life," Commoner wrote, converting the earth's "endless cycles into manmade linear events."

In June 1987 Commoner angered some enviromentalists, already frustrated by the antienvironmental policies of the Reagan administration, when he wrote a lengthy and critical assessment of the nation's environmental progress in *The New Yorker.* In it he argued that although there has been modest improvement in cleaning up the environment, our successes have occurred only when the "relevant technologies of production" have been changed "to eliminate the pollutant. If no such change is made, pollution continues unabated or, at best—if a control device is used—is only slightly reduced." In essence, he claimed that our efforts to control environmental pollution by means of governmental regulation have not been very successful. The article is reprinted here in its entirety.

Michael Clark, president of the Environmental Policy Institute in Washington, D.C., was the first of a number of friends to suggest that the issues dealt with in the *Amicus* series and the Commoner article be expanded into a book. The logical publisher was Island Press, which in a few short years has played an invaluable role as a publisher and distributor of semitechnical environmental works. Its recent titles include *The Report of the President's Commission on Americans Outdoors* (published over the objections of the White House), *Last Stand of the Red Spruce* (with NRDC), and *The Forest and the Trees,* as well as the first three works in a series of conservation classics: Gifford Pinchot's *Breaking New Ground,* Edward Faulkner's *Plowman's Folly* and *A Second Look* (together in one volume), and J. Russell Smith's *Tree Crops: A Permanent Agriculture.*

In 1986, Island Press also published an edition of *An Environmental Agenda for the Future,* written by the leaders of ten national environmental organizations and edited by Robert Cahn. *Agenda* was intended to be a consensus statement and as such attempted to define the issues on which

most environmentalists could agree and to suggest ways of dealing with
them.

This book, *Crossroads: Environmental Priorities for the Future,* is
intended to be more of a three-panel mosaic. Between the interviews and
analysis of Part I and the more detailed statements of Part II, we have tried
to define the cause of environmentalism and to construct an image of the
movement—past, present, and future.

For those who think in terms of stereotypes or who believe that the
movement is a monolithic special interest, read on. Environmentalists are
a remarkably diverse sort, which accounts for the strength of the movement
(and perhaps its monolithic image). At one point we considered borrowing
from William James and calling this book *Varieties of Environmental
Experience.*

We asked our contributors—all veteran activists—to consider where
we stand in terms of the environmental goals and objectives set over the
past twenty years. Why? Because history teaches us that social progress is
achieved only by those who recognize change and respond to it. The
fundamental causes of environmental destruction have changed very little,
but our knowledge of the world around us and the means of effecting
change at our disposal have multiplied considerably in less than a generation. It is time to consider the course of the second generation. Specifically, we asked:

—Have the environmental reforms set into motion in the 1960s and
'70s been effective in cleaning up the environment?

—Is today's environmental movement responsive to public concerns
about environmental deterioration at local and regional levels?

—On the basis of current trends, will the level of environmental
protection achieved by the turn of the century be sufficient to prevent
significant changes in the biosphere?

—Have the conservation and environmental movements had a significant influence on our relationship with nature?

This last question reminds me of the gag line, "Have you stopped
beating your wife?" If you have not, you should. If you have, you should
not have beaten her in the first place. In both cases it sounds like a bad
marriage, an apt description of our relationship with nature.

Following Earth Day in 1970, there was great hope that by defending the environment by writ we would drive technology toward more benign means of achieving our economic objectives. There was not much talk of changing our objectives. The political assumption was that we could have more and do less harm. The thought that such a proposition might be fraught with greed and global inequity or ecological impossibility is only now beginning to be understood. We are at a crossroads that requires more of us than we have ever given. This time the challenge is neither to subdue nature nor to protect it from ourselves but to act as both a *part* of nature and a *member* of a global community. As many of our contributors note, if there is not also an emotional and spiritual commitment to these objectives, enlightened policy and politics will be of little avail.

Peter Borrelli
Vischer Ferry, New York
July 1988

Part I

Trends

1

Environmentalism at a Crossroads

Peter Borrelli

ONE OF THE BY-PRODUCTS of the Reagan era has been a rare degree of introspection among environmentalists. Obstacles placed in the way of aggressive environmental action—federal budget cuts, legislative stone-walling, and appointments committed to undoing the progressive gains of the sixties and seventies—have forced environmental advocates not only to try harder but to reexamine priorities and strategies.

For those who enlisted in the movement on Earth Day, and especially for those who already were active during the preceding decade, the Reagan era has been doubly difficult to cope with. During this period, some environmentalists have begun to feel that the idealism of the movement is mysteriously expiring. And the belief that environmental programs had been woven into the fabric of American society has been shaken, even shattered.

Reflection is proving to be both stabilizing and divisive. On the one hand, there is a growing consensus about priorities. On the other, the debate over tactics and management has torn apart some organizations and led to the proliferation of smaller, local organizations. Recently, *The Amicus Journal* asked nearly one hundred environmentalists around the country to reflect on the past, present, and future directions of the movement.

Every April, Gaylord Nelson, chairman of the Wilderness Society, is asked whatever happened to Earth Day. Where's the razzmatazz, the crowd, the sense of purpose that attended the birth of the modern environmental movement?

The former U.S. senator from Wisconsin, who authored the legislation that proclaimed April 22, 1970, Earth Day, usually returns an understanding smile and patiently explains that the media hype that occasioned the nationwide event was the means, not the end, of environmentalism. "Earth Day did not create the interest; all it gave was an opportunity to express itself. My sole purpose was to force the issue on the politicians and to make it a part of the national dialogue."

During a fourteen-year career in state government, he had witnessed a deep-seated interest in the outdoors. Hunting, fishing, and camping were central to Wisconsin's good life, and those who enjoyed it had begun to realize that the landscape of their youth was changing for the worse, that their love of country was growing in direct proportion to their hatred for the city, which often was the source of the countryside's decline. Nationally, writers such as Rachel Carson, Aldo Leopold, and Paul Sears were alerting a growing army of citizens to the ravages and dangers of postwar growth and technology.

Despite this nascent concern, the conservation movement was still largely a movement of nature lovers who joined in the National Audubon Society's Christmas bird count, hiked with the Sierra Club, or fished with the Izaak Walton League. Earth Day had a metamorphic and democratizing effect. Both strains of the American conservation movement—the pragmatic reform tradition established by Teddy Roosevelt and Gifford Pinchot, and the preservationist tradition established by Henry David Thoreau and John Muir—expanded their agendas and took on new members by the thousands. The common denominator was pollution, and for a short time there was a merger of the political left and right, radicals and reformers, professionals and volunteers. The outcome was one of the major social movements of the twentieth century.

Recalling his arrival in Washington, D.C., in 1963, Nelson says, "There were not more than five broad-gauged environmentalists in the Senate: Lee Metcalf, Ed Muskie, Frank Church, Clinton Anderson, Hubert Humphrey." When he introduced a ban on DDT during his freshman year, John Dingell (D-Michigan) was the only representative willing to sponsor a companion bill in the House. By comparison, today he believes that "all one hundred senators would claim to be environmentalists, and at least fifty to sixty are very knowledgeable about one or two issues. I knew we had arrived last year when during a floor debate in the Senate over the administration's proposal to sell off some public lands, Jesse Helms (R-North Carolina) stood up and said, 'This goofy administration is not going to get away with selling national forest lands.'

"If you had told me [in 1970] that Congress would authorize $18 *billion* for waste water treatment in 1986, I would not have believed it. The last Johnson budget only had $250 million in it."

From Nelson's perspective, the movement is growing exponentially. "The reason groups like the Wilderness Society, National Wildlife Federation, and NRDC have clout today is that the base [of environmentalism] is broadening. Senator Proxmire calls the environmental lobby the most effective one in Washington." Not, he adds, because groups like the Wilderness Society have grown (from 37,000 members and a budget of $1.5 million in 1981 to 165,000 members and a budget of $8 million today), but because the movement has expanded at all levels. "The movement is as vital as ever," says Nelson. "The trends are all positive. It's all part of a big mosaic."

If Nelson is one of the graybeards of the modern environmental movement, Denis Hayes must be considered one of its middle-aged veterans. The founder and organizer of Earth Day, Hayes later became head of the short-lived Solar Energy Research Institute, established during the height of the energy crisis and abolished early on by the Reagan administration. Today, he practices law in Palo Alto, California, and chairs the Washington, D.C.-based Fund for Renewable Energy and the Environment (successor to the once vital, now defunct Solar Lobby).

"In terms of sustainability, the movement has exceeded our expectations," says Hayes. "It's managed to avoid the American tendency to come and go like hula hoops, Davy Crockett hats, and punk rock haircuts." He contrasts it with the nuclear freeze movement, which recently merged with SANE: white hot in '80, torpid in '87. "There was a conscious decision in organizing Earth Day that [environmentalism] would not be posited in a fashion that was ideologically exclusive; there was room for middle-class housewives, business executives, radical college kids. Its capacity to reach out to an enormously broad set of constituents, and to give people a way to assimilate the values in things that they can consciously do and affect—and see consequences, have given it staying power."

Doug Scott was the student organizer of Earth Day at the University of Michigan at Ann Arbor. A wilderness advocate, he was a point man in the new old brigade made up of young people primarily interested in carrying on the work of Robert Marshall, Howard Zahniser, and Horace Albright. (The label "environmentalist" never quite wore as well on them as it did on the Columbia students protesting Con Ed emissions in New York—the vanguard of a talented and ambitious new new movement that carried on its advocacy with such groups as Nader's Raiders, Environmental Action, the Environmental Defense Fund, Natural Resources Defense Council, and Friends of the Earth.) Today, he is the conservation director of the Sierra Club, which claims a membership of about 400,000 and a budget of $23 million.

"Our ambitions, plans, and visions about what was the art of the possible for preserving wilderness were feeble compared to today's reality. Eighty million of the 91 million acres in the wilderness system have been added since the Wilderness Act was passed (1964). This is an enormous achievement, and most of that land was never on anybody's agenda."

Scott recalls that in the late sixties, Michael McCloskey, then conservation director (now chairman) of the Sierra Club, wrote a letter to the Forest Service and made the mistake of enclosing a map of the club's "ultimate" plan for the Siskiyou wilderness area in northern California. "Now it's three times as large and the Forest Service keeps hauling out this yellowed letter and saying, 'but you guys said.' Our fondest ambitions have underestimated the growth of our political clout."

Earth Day did much more than attract hot new blood to the established conservation movement of the fifties and sixties. A whole new agenda, focused on "the quality of life" (a phrase worn thin during the seventies and seldom heard today), poured out of Washington. The new environmentalists were highly legalistic and advocated clean air and water as matters of right. Indeed, some of the new movement's leaders such as William Futrell, a former law professor in Georgia (now president of the Environmental Law Institute in Washington, D.C.), had cut their activist teeth in the civil rights movement. A few others like John Adams, NRDC's first and only executive director, were crime fighters at the U.S. Attorney's office in Manhattan. In the context of the times, pollution was viewed as an injustice; a logical extension of the disdain so powerfully expressed by Teddy Roosevelt for those who robbed and wasted the country's natural resources.

With the passage of the Clean Air Act of 1963, Congress set in motion a nationwide program to achieve acceptable air quality. But it was not until 1970—at the height of public attention—that Congress further required achievement of national air quality standards to protect human health by 1975. (Amendments in 1977 extended the deadline for the attainment of ambient air quality standards in most areas to 1982.)

Under the Clear Water Act (1972), the government was directed "to restore and maintain the chemical, physical, and biological integrity of the nation's waters," especially to eliminate the discharge of pollutants by 1985. And so it was with a panoply of environmental concerns, as legislation was passed to protect drinking water, control hazardous wastes, regulate toxic chemicals, prevent ocean dumping, and reclaim strip mines.

The rush of legislative reforms spawned a whole new field of law, which in turn led to landmark litigation the names of which—Storm King, Calvert Cliffs, Overton Park, Vermont Yankee—roll off the tongues of early environmental attorneys such as David Sive of New York (a board member at one time or another of virtually every major conservation or environmental group) like the names of famous military campaigns.

Russell Train, the dean of Washington environmentalists, chaired the first Council on Environmental Quality (CEQ) under Nixon and now presides over the recently merged World Wildlife Fund-U.S. and Conservation Foundation. He recalls with a mixture of bemusement and pride that the movement of the sixties grew "without any apparent divine guidance or highly visible leadership. We just started identifying problems and fussing around with them and coming up with a program, which typically was a regulatory program. By and large, given the agenda of the time, we did pretty damn well."

Train was astonished by a recent scathing attack by Edith Efron, author of *The Apocalyptics: Cancer and the Big Lie,* accusing him and other Earth Day adventists of manipulating public opinion on toxic substances. Referring to his speech before the National Press Club in 1976, Efron claimed that "even [Samuel] Epstein and [Ralph] Nader were made to look diffident and uncertain by Russell E. Train who told his audience that 'because of toxic chemicals in use in industry, all Americans' lives were in peril.' He said the nation must stress 'the prevention, rather than the treatment of disease' and that such industrial dangers must be dealt with before, not after they entered the environment." Efron added that Train's "historically significant remarks" were incorporated in the Toxic Substances Control Act and contributed to its passage.

Train laughs at the thought that such a statement could have been considered "extremist rhetoric" then or now. "Philosophically, temperamentally," says Train, "I'm sort of flopping around in the middle. Dave Brower once joked, in retaliation for a remark I had made about his making others appear reasonable, 'Thank God for Russell Train for he makes it possible for almost everyone else to sound outrageous.' "

The story illustrates Hayes's observation on the ideological inclusiveness of the movement in the late sixties and early seventies. Today, however, there is evidence of increasing diversity as ideological lines are being drawn, partly to set new directions, partly to reassess the past.

Although the Reagan administration set the stage, several incidents within the movement focused attention on strategic, ideological, and organizational differences. The first was the open and bitter breakup in 1985 of Friends of the Earth (FOE) over managerial and tactical issues. It resulted in the forced resignation of its charismatic founder and chairman, David Brower, perhaps the one person in the country who came closest to embodying the values of the national movement (Sive often has likened Brower to Martin Luther King, Jr.).

At about the same time, a number of other national organizations—

including the Sierra Club, National Audubon Society, Environmental Defense Fund (EDF), and the Wilderness Society—underwent major reorganization in their front offices.

The movement, which had been expanding its international activities, also was stunned by the French government's sinking of the Greenpeace flagship, *Rainbow Warrior*, and the killing of a crew member. Never before had the movement come up against such violent opposition.

And finally, the publication of an innocuous sounding report by the heads of ten major environmental groups, entitled *An Environmental Agenda for the Future*, prompted regional environmentalists to voice ideological independence from the nationals.

In Washington, San Francisco, and New York, the corporate homes of most of the national organizations, there is a great deal of talk these days about the grass roots. For older groups like the National Audubon Society and the Sierra Club, citizen action is where it all began. For groups like NRDC and EDF, which are supported by growing memberships but do not have local affiliations, the grass roots, nevertheless, are where much of the action is. The reason: toxic substances in the air we breathe, water we drink, food we eat.

Major segments of the movement have experienced both a definitional and substantive shift from the earlier, generic quality-of-life agenda to a human health and well-being agenda. While the two are not mutually exclusive, the latter is more personal and touches the lives of more Americans than virtually any other social issue, save the economy and nuclear war. *The New Republic* calls today's (the new new) environmentalism "America's issue."

The administration's policies, according to the magazine's editors, "are based on one of Reaganism's pet conceits: that environmentalists are a fringe group of anti-growth elitists. What Middle America really wants are jobs and industrial growth, however many poisonous byproducts that growth may generate. . . . In fact, Reagan's ideology indisputably places *him* on the fringe." His appointments of James Watt as interior secretary and Anne Gorsuch Burford as administrator of EPA demonstrated how totally out of touch he is with the American people on environmental issues. And his pocket veto of the Clean Water Act last fall showed how much out of step he is with Congress. (The water bill was Congress's first order of business in 1987, and again it passed by a sufficiently wide margin to override a second veto.)

Ironically, the antienvironmental policies of the Reagan administration, most hard felt by labor and minorities, have strengthened the alliance between these groups and the environmental movement. Today, pollution is treated more as a people problem. Generic concerns about industrial policy and standards of emission and discharge have yielded to concerns about the health and safety of migrant farmhands, industrial workers, and inner-city children.

The most striking characteristic of the new new movement is that most of its followers are not card-carrying environmentalists. The national groups have benefited handsomely during the past five or six years; their ranks have swelled from about 4 million in 1980 to about 7 million in 1987. Some critics argue that these figures are misleading, because there is some overlap of membership among the groups (perhaps as much as 30 percent). On the other hand, if each group were to include lapsed members and put a number on what it believes to be its numerical base of support, it is conceivable that the nationals have a following of 10–15 million.

If the growth among the nationals has been robust, the growth at the local level has been explosive. Estimates are crude and varied, but there may be as many as 25 million people involved in one way or another in local and regional issues ranging from the hot new concerns about groundwater contamination, nuclear waste disposal, and municipal landfills, to traditional conservation and preservation issues involving wildlife and public lands.

Lois Gibbs, who first organized her neighbors to protest conditions at Love Canal in Buffalo and now heads the Citizen's Clearinghouse for Hazardous Wastes in Washington, works with 1,300 local groups, most of which did not exist three years ago. The threat of chemical contamination, combined with general frustration over the pace of governmental action (and to some extent, the remoteness of national environmental organizations), has been a powerful catalyst in their formation.

For example, widespread concern over the uses and abuses of lawn-care chemicals has led to the creation of dozens of groups like Karen Blake's Help Eliminate Lawn Pesticides (HELP).

As Doug Scott observes, the community-based toxics groups have almost no formal connection with the national conservation organizations and "don't much need us." They especially do not want to get "swept up in our institutions, with all the internal red tape. And that's the way it ought to be. The grass-roots reality is very big, damn close to lawless, and inherently frustrating to deal with if you have a neat mind and want everything well organized."

The premier example of the power of these new grass roots was the passage in California in 1986 of Proposition 65, the get-tough-on-toxics initiative that pitted citizen activists against oil and chemical companies and the Deukmejian administration. The initiative won overwhelming voter approval.

Tom Hayden, the former head of Students for a Democratic Society and anti-war activist, is today the California state assemblyman who chaired the campaign to adopt Proposition 65. He was first drawn to the issue when a close friend died of leukemia in the mid-seventies. Prevention, he concluded, required minimizing exposure to chemicals. He and others persuaded then-governor Jerry Brown to adopt a far-reaching state toxics program that was a national model until toxics politics got in the way. "The toxics program in California and at the national level," says Hayden, "has become a case study of government throwing money at a problem without a serious plan to resolve it." The California program is currently under investigation by both the EPA and FBI over alleged illegal contract procedures.

The impetus behind Proposition 65 was a widely held public belief that the government's approach to toxics relied too heavily on technical fixes instead of prevention. Aimed at prohibiting contamination of drinking water with chemicals known to cause cancer or reproductive effects, Proposition 65 provides stiff penalties and a mechanism for citizen enforcement. Hayden immodestly calls it "the biggest environmental victory ever" and views it as the beginning of a nationwide crusade. People are simply fed up with government's handling of the problem: not just with Reaganism, but with the entire regulatory approach, which has led to a string of pitched battles with industry over one chemical after another.

But trend is not destiny. California environmentalists had not tasted victory in a dozen previous attempts at initiatives since 1972, when voters approved the state's revolutionary coastal protection plan. The difference with Proposition 65 was a well-financed ($1.7 million), sophisticated political campaign. And with the help of Hayden's wife, actress Jane Fonda, the stars came out—Whoopi Goldberg, Michael J. Fox, Linda Evans, Chevy Chase, Morgan Fairchild, and John Forsythe, to name a few.

The concern of many environmental leaders is whether massive public awareness leads automatically to progressive change. Lucy Blake, executive director of the California League of Conservation Voters (the largest state environmental political action committee), thinks not. "We're facing a real challenge, because people are operating on such a superficial level in their personal relationships, and certainly in their organizational relationships—by writing a check, for example, which says 'I support you, send me your mail,' and in terms of their voting. The environmental movement can very easily be coopted by people spouting platitudes about the environment, but who don't really mean it and don't deliver."

The national organizations play an important role in the overall scheme of things, but she fears that in the lengthy process of their winning acceptance of environmental values, they have lost their bite and unwittingly contributed to public apathy. "You need to create anger and outrage to get people involved. You need to be able to tell people that they don't have to live like this, breathing brown air all the time."

According to Blake, environmental activism is most effective at the local level among "people discovering they have shared problems: water contamination, traffic, development." The national groups could not address all the local needs if they tried, although, she adds, they can provide much needed training and technical support.

Her remarks echo the complaint that the nationals have become bureaucratic and remote, that strategic planning has shifted from the kitchen tables and basement playrooms of volunteers to glittery Washington, D.C., offices. To some extent the breakup of Friends of the Earth was over the issue whether to give in to the lure of Washington. Brower and many members feared that FOE would lose its identity and moral edge; they said no to a move to Washington. The more numerous critics, for whom capital politics had become *the* issue (FOE endorsed Mondale for president in 1984 even before he was nominated by the Democrats), said yes, and they went.

The discomfort with Washington may stem from the sense that environmentalism is a social movement, not a special interest. In 1985, when Peter A. A. Berle took over as president of the National Audubon Society (which together with NRDC and EDF still maintains its headquarters in New York), he was faced with a proposal to move to Washington. Aside from its being frightfully expensive (Audubon has a New York staff of about 100), Berle's sense was that in D.C., Audubon might appear to be "just another trade association."

Identity remains a central concern of most groups, but the inescapable reality of today's movement is that what goes on in Washington is at least as important as what goes on elsewhere. The difference may be that for all the glamor of the nation's capital, the day-to-day work of professional environmentalists lacks the excitement of a popular cause.

For the past ten years—even before Ronald Reagan stormed the White House—the task of environmental groups has been the tedious grind of dogging federal agencies charged with implementing the environmental laws enacted in the sixties and early seventies and of preventing Congress from backsliding. It was difficult enough during the Carter administration, which was favorably disposed toward environmental protection and conservation. The Reagan administration's policies have simply meant that more time, energy, and money had to be spent making the system work. The appointment of James Watt as secretary of the interior was a bizarre stroke of good fortune. The archetypal villain on the one hand made things a bit easier. The nationals churned out more than 100 million appeal letters during his brief tenure, in some instances more than doubling their membership. On the other hand, it proved costly and quite conceivably did little to advance the cause.

The paradox is that the nationals are now being criticized for being too big and having achieved some measure of what can only be called success. Brower cites, for example, the Sierra Club's 59 chapters, 339 groups, and 60 standing committees and complains, "It's becoming like Velveeta; everything must be processed." Others, like Lucy Blake, believe that the process, though studied, is highly democratic. When the club does take action, "People move together and don't drop off. The Sierra Club is the Girl Scouts of the American conservation movement. It's great because it is easy to relate to."

The pros and cons of bigness are the subject of similar discussions and debates within virtually all the major organizations that have local chapters and groups. With bigness have come greater power and the ability to effect change, but at the price of familiarity.

The frustration over the size and complexity of the movement cuts both ways. While the grass roots feel cut off at times, staffers with the big organizations often complain of being lost in the crowd. Doug Scott recalls that ten years ago "you could have meetings of the environmental movement in Washington to talk strategy about Alaska and everybody was there and everybody knew everybody else. When I go to meetings today, half the groups aren't there and everybody's a stranger.

Small may be beautiful, but most environmentalists who started out ten, fifteen, or more years ago feel that bigger is better. The environment has become a central concern of government, business, and the public consciousness. As Russell Train says, "Environmental concern has been institutionalized." In less than a generation, the United States has constructed an environmental program unequaled in the world. The job of making it work is often time-consuming and highly technical. At times it may seen uninspired and gray.

When the Group of Ten,* a newly formed coalition comprised of the heads of ten major environmental organizations, issued *An Environmental Agenda for the Future* in 1985, it was criticized for being what it was: a consensus statement on future environmental issues, as seen through the eyes of those most closely involved with existing environmental institutions. "In their reflexive adherence to legislative solutions," wrote Ed Marston in the highly respected *High Country News,* "they sound like chemical companies responding to increased resistance of pests to pesticides by prescribing ever larger doses of ever more potent poisons."

Although the Group of Ten highlighted its concern for a more global

---

*The Group of Ten is an informal group made up of the chief executive officers of the Environmental Defense Fund, Environmental Policy Institute, Friends of the Earth, Izaak Walton League of America, Natural Resources Defense Council, National Audubon Society, National Parks and Conservation Association, National Wildlife Federation, Sierra Club, and the Wilderness Society.

view ["Only by understanding these global issues and giving them their necessary place on the scale of priorities can citizens improve opportunities for sustainable human progress and preservation of the earth's environment"], Marston was hoping for something "more far-seeing, more imaginative" on the home front to deal with the unfinished domestic agenda of the past decade. Marston seems wary of the global trend among the major environmental organizations. It coincides with a resurgence of interest in traditional conservation issues such as forest practices in the Northwest and wildlife protection (being picked up by groups like Earth First! and the Sea Shepherds Conservation Society) and the emergence of the toxics groups and communities organized around the principles of bioregionalism and Green politics. Within each of these camps there is a great interest in personal involvement, something the nationals cannot always offer. By their very nature, international issues preclude direct involvement.

In their criticism of the nationals the new groups lack a sense of history. They tend to speak as though environmentalism began on Earth Day and has continued on more or less a single frequency ever since. The movement, of course, was quite well developed long before Earth Day; and more importantly, there has been constant commingling of old and new issues and strategies since at least the time of Muir's and Pinchot's famous quarrel. Newcomers to the movement (whether new new or new old) often complain that the nationals do not speak with a single, crusading voice, overlooking the fact that even at "the top" there are conflicting agendas and personalities. Witness the absence of representatives from Greenpeace (with its 800,000 supporters) and Defenders of Wildlife (with 80,000 members) among the Group of Ten.

Another frequent criticism, which the press has been quick to pick up, is that the movement is becoming politically and temperamentally more conservative. There were reported shake-ups at the Wilderness Society, EDF, National Audubon Society, Sierra Club, FOE, and Greenpeace, all within a period of about two years. The story ran that the movement had caved in to Reaganism and turned to leaders armed with MBAs rather than terrible swift swords. It made good copy, but belied many of the facts.

The early eighties represented a period of rapid growth and challenge for a number of the nationals. The lesser known fact is that most incurred deficits as they pulled out all the stops in an effort to stop Watt. Some (like FOE, Environmental Policy Institute [EPI], and EDF) tottered, while virtually everyone (including NRDC) had difficulty meeting their budgets. It was a period when no one felt he or she could leave the helm. According to Mike McCloskey, who had been itching to step down as the Sierra Club's executive director (a position he held from 1969 to 1985), "There might have been a more gradual turnover during the early eighties but for the Reagan administration. Some people stayed longer, because it was an exciting fire fight and they wanted to see it through."

By 1983, however, it was evident that the fight would not be over soon and that if internal changes were needed, there was no advantage in waiting. Contrary to reports, the groups did not turn to business corporations for their new leaders. Fred Krupp, who took over EDF, was recruited from a public-interest law firm in Connecticut modeled after NRDC. Peter Berle, who succeeded Russell Peterson at Audubon, was a nationally known environmental commissioner of New York State. Douglas Wheeler, who succeeded McCloskey, had headed the American Farmland Trust in Washington, D.C. Steve Sawyer became executive director of Greenpeace after having served on its five-member international governing board (he was skipper of the *Rainbow Warrior* during its fateful voyage). George Frampton, a former Watergate attorney, took over from William Turnage at the Wilderness Society. Joyce Kelly left the President's Commission on Americans Outdoors to succeed Allen Smith as head of Defenders of Wildlife. And with the exception of Turnage, who moved to England in pursuit of a new career, virtually all of the outgoing class stayed with the movement. Peterson is now president of a new international organization called The Better World Society; McCloskey is chairman of the Sierra Club; Brower is chairman of Earth Island Institute; Smith is vice president of the Wilderness Society.

The one noteworthy trend that went largely unnoticed but which troubles some environmentalists was that most of the organizations did not hire from within. Career development is not a strong suit of the movement. Allen Smith, who started out in the sixties as a Sierra Club volunteer in Boston, mourns the loss of what he calls the "community culture" of the movement and complains that "the community isn't growing its top leadership."

Several years ago, the Conservation Foundation recognized the problem and surveyed environmental organizations of all sizes to determine what if anything could be done to strengthen management skills. It agreed with Smith's assessment and recommended that groups institute management training programs for both staff and grass-roots activists to cope with the demands of a growing movement.

Whereas in the early seventies the nationals might have run one major campaign at a time, today Audubon, Greenpeace, or NRDC may have half a dozen campaigns with relatively equal priority running simultaneously. As UCLA environmental planners Margaret FitzSimmons and Robert Gottlieb put it, when the organizations became busier and more complex, "computerization of mailing lists took precedence—in fact substituted for—community organizing; the skills of the lobbyist, litigator, and expert were valued over the passion of the outraged housewife, the angry consumer, or even those who felt bereft by the deterioration of rivers, streams, and mountains."

Today the relationship between the nationals and state and regional organizations is more likely to be contractual than inspirational. Local groups look to the nationals, not for great speeches and ideas, but for experience and expertise on federal legislation and programs. For example, NRDC's water-pollution enforcement program (the E-Team) has been working with local and regional groups in a dozen states, suing several hundred industrial polluters under the citizen suit provision of the Clean Water Act. This past spring NRDC and the Chesapeake Bay Foundation reached an historic out-of-court settlement with Bethlehem Steel in connection with the company's alleged violations of the Clean Water Act at its Sparrows Point, Maryland, plant. The settlement was the largest ever won in a citizen suit.

National Audubon also has gone local by launching a multi-million-dollar education program in several hundred school districts, an effort it expects to double annually for the next several years. The Environmental

Policy Institute runs an activists' training program in Appalachia on federal and state strip-mining enforcement. The Environmental Law Institute publishes a national newsletter on wetlands, serves as a clearinghouse, and provides a technical backup for hundreds of state and local groups battling developers and the Army Corps of Engineers.

Far from being isolated from the grass roots, the nationals are being overwhelmed by requests for local assistance. The real difference may be that ten or fifteen years ago, "the word" would emanate from Washington that a particular law was needed and groups around the country would organize around the issue; it was more or less a top-down agenda. Today, often as not, the flow is reversed; the action is bottom-up, since it is at the local level that laws and programs set in place over the past two decades are implemented. The shift was well under way before Ronald Reagan arrived in Washington. For example, during the first years of the Carter administration, the Environmental Protection Agency began delegating enforcement authority to the states under state-EPA agreements. By cutting EPA's budget, the Reagan administration accelerated the process whether the states agreed to it or not.

Separate and apart from the inevitable institutionalization and partial bureaucratization of the movement, there remains the nagging question whether it has in fact become more politically conservative. Conversely, one might ask if there ever was a time when it was more liberal. There appears to be no simple answer, in part, as Denis Hayes suggests, because political and ideological differences were subordinated at the time of Earth Day. The movement in 1970 pulled together a broad political spectrum of ultraconservative hunters and fishermen with little or no action agenda; pragmatic reformers in the mold of Teddy Roosevelt and Gifford Pinchot; middle-class moderates newly awakened to the dangers of pesticides and ravages of industrial pollution; hard-line preservationists carrying on the anti-industrial tradition of John Muir; and elements of the old and new Left for whom the environmental crisis was positive proof of the need for a new social order.

There was remarkably little tension between and among activists throughout the seventies. (It helped that for the first six years there was a Republican administration in Washington.) The agenda was so crowded and diverse that cooperation was a matter of survival. The coalition that formed to block development of a U.S. supersonic transport (the SST) was virtually identical to the one that defeated a crass effort to open the national forests to accelerated logging by authority of a National Timber Supply Act.

But by 1980, there was less of a sense of common cause. Organizations had begun to concentrate their energies and to specialize on specific statutes, programs, and agencies—often determined by members' and supporters' expectations, house expertise, or available funding. The appointment of a number of environmental activists to high posts in the Carter administration (despite the environmental gains of the early seventies, few activists had risen to positions of high governmental responsibility) also had the effect of identifying the new environmental agenda with an antiestablishment wing of the Democratic Party. The image was reinforced by the emergence of a strong and aggressive environmental program under California Governor Jerry Brown.

Phillip Berry, an Oakland attorney and president of the Sierra Club from 1969 to 1971, believes the concern about creeping conservatism is really a concern about "moral slacking off by leaders of the movement, of their going over to the *other side,* forgoing principles. . . . But I don't think it's happening in that sense. It was easy in 1970 to go around pointing out problems nobody else had seen and to appear mildly profound. There may be an appearance of more caution, a slower approach today, but that's because the complexity of the issues requires a great deal more sophistication. It's not because we've become conservative like Ronald Reagan. Besides, he's an utter radical; we're the true conservatives."

McCloskey, perhaps the movement's most inveterate politician, says, "The correlations between political philosophies and environmental commitments have never been clear. Some see old wise-use conservationists [Pinchot] as conservatives, while the new environmentalists are to the left; and they see the age 30–50 population cohort as the heart of support for both new liberalism and the environmental movement. On the other hand, some public opinion surveys show little difference in the degree of support for environmentalism among both liberals and conservatives. Aspects of environmentalism appeal to both camps. Liberals favor government intervention via regulation and resist business interests. Conservatives favor reduced outlays for public works and an end to subsidies. It has also been pointed out that radical environmentalists diverge from the normal left over issues of income distribution, decentralization, and abortion."

While right/left distinctions may not be very instructive in predicting positions on issues, surveys show that about 60 percent of the members of environmental organizations consider themselves to be liberal or to the left of center. Environment Action appears to be the most liberal, while the National Wildlife Federation, whose profile most nearly matches that of the general public, appears to be the most conservative. But what may be more revealing is that the political extremes are numerically small among all of the nationals.

A distinction that McCloskey takes more seriously is between what he calls norm- and value-oriented environmentalists. The norm-oriented enviros are reformers who do not quarrel with the basic ways in which society operates. The value-oriented want to change the relationship of individuals to society and the ways in which society works. He considers most of today's major national organizations to be norm-oriented. The distinction, adds McCloskey, "turns on whether it is wise to work within the context of the basic social, political, and economic institutions to achieve stepwise progress, or whether prime energies must be directed at changing institutions."

For the most part, the national organizations downplay their inner-most ideological orientations. They have become strategically deft at carrying out their agenda; lobbying Congress one day, going to court the next, and meeting with labor and religious leaders the next. Greenpeace is a good example. Most of its supporters would never think of shinnying up a smokestack or playing chicken with a supertanker in an inflatable dinghy, but they approve of the tactics' publicity value. Less visible but critically important is the follow-up. The group is universally admired for its persistent and effective appearances before the International Whaling Commission.

Similarly, NRDC's or Sierra Club's skill at blending into the Washington landscape contrasts sharply with their high visibility in court. In 1986, NRDC filed nearly three dozen lawsuits. The Sierra Club Legal Defense Fund added another twenty-one. The numbers belie the notion bandied about a great deal last year that the nationals were adopting a new strategy of compromise evidenced by their willingness to sit down with industry and government. The clearly observable trend is that industry now seems more willing to sit down with environmentalists. "Industry knows that the new laws and legal victories won by environmentalists over the last decade aren't going away," says William Reilly, president of both the Conservation Foundation and World Wildlife Fund-U.S. (WWF). While the antienvironmental policies of the Reagan administration held out hope for some industrial interests in the early eighties that good times were ahead, public and political support for the environment has not been untracked. Realization of this, combined with some genuine examples of enlightenment within the corporate world, has led to some dramatic turnarounds on the part of electric utilities and open negotiations with oil companies over offshore exploration. Virtually every major environmental group today can draw upon the expertise of scientists, economists, attorneys, and planners to shape a better project or statute. If there is a new or "third wave," as Fred Krupp, executive director of EDF, has described it, it seems more a function of the movement's increased leverage, its ability to forge instant alliances with other social movements, and additional resources rather than any profound shift in thinking.

The ability of the nationals to tailor their strategy to fit a dozen or more priorities at a time contrasts with some of the newer groups' narrower agendas. Earth First!, based in Tucson, and the Sea Shepherd Conservation Society of Los Angeles are the two most highly publicized groups to break ranks with the national movement. They represent the activist brigade of a growing network of local movements (a reemergence of the sixties New Left that espouses such causes and ideas as Green politics, deep ecology, and bioregionalism). The two groups practice environmental sabotage, or "ecotage," and have claimed responsibility for spiking trees, lying down in front of bulldozers, and sinking whaling ships. They employ these tactics to draw attention to such issues as the government's cutting of old-growth timber and road development in national forests. Ironically, while they talk about the need for radical social change, most of their actions have dealt with relatively narrow and traditional conservation issues relating to animal rights and abuses of public lands.

Earth First! founder Dave Foreman, a former lobbyist for the Wilderness Society, says, "We're all for what the Sierra Club, the Natural Resources Defense Council, and Audubon have been doing. We'd like to see them take a little stronger stand, be more strategic and more philosophical, and understand why we're trying to do this. But we also need more people doing direct action, stepping outside the system and coming up with conditions of what an environmentally sane world should be. We need people 'monkeywrenching.' "

While monkeywrenching is unlikely to attract a great many people, the message of Earth First! is not much different from that of many environmental groups examining global issues. "We're in the middle of the greatest biological catastrophe since the dinosaurs died off 60 million years ago," says Foreman. "Won't anybody wake up?"

Lester Brown, president of the Worldwatch Institute in Washington, D.C., is among those sounding the global alarm. His group publishes exhaustive studies on the state of natural resources, especially in the third world. A recipient of a MacArthur Foundation "genius" fellowship, Brown recalls, as though it were a century ago, the 1972 United Nations Conference on the Human Environment held in Stockholm. It was organized by the industrial countries and the emphasis was primarily on pollution. Today's concerns about global warming, deforestation, ozone depletion, soil erosion, and world food production, he says, represent a profound shift in public awareness and thinking. "We are in for a very difficult time," but like most U.S. environmentalists looking at the global agenda, there is optimism in his voice.

International concerns have spawned a number of new groups, including the MacArthur Foundation-funded World Resources Institute headed by former CEQ chairman Gus Speth, the Better World Society, and the Global Tomorrow Coalition. A close look at the makeup of these groups reveals direct ties with some of the major environmental groups, most notably NRDC, National Audubon, and WWF. The Global Tomorrow Coalition includes a wide range of institutions, including many of the mainstream environmental groups. The coalition was an outgrowth of the Carter administration's 1979 *Global 2000* report, an unprecedented interagency examination of the state of world resources.

Far from being a duplication of organizational effort, the new groups have made it possible for the environmental movement to expand its agenda with relative ease—so much so that the historical significance of what has happened has gone largely unobserved. In the course of the last five to seven years, U.S. environmentalism has metamorphosed from a largely domestic, even chauvinistic social movement to a global crusade with much of the missionary zeal of the Peace Corps.

Among the internationalists there are two compatible camps: the survivalists and the humanitarians. The survivalists are interested in what the late Barbara Ward called "planetary housekeeping." "There are basic principles that can define the relationship of human beings to the biosphere," says New York State Commissioner of Environmental Conservation Thomas Jorling. "If we can comply with those basic principles, we can achieve a sustainable future. Most importantly, we must avoid changing the basic geophysical processes of the biosphere that allow us to inhabit the planet. That's our objective and we can meet it. This is a very optimistic view of the future."

The humanitarians share this planetary concern, but focus their attention on the third world where they are promoting the concept of sustainable development. Again, Barbara Ward described the challenge well: "We have reached the point of talking together about the great common tasks of humanity—preserving our living environment, feeding the hungry, giving shelter to all our fellow creatures, treating with greater care and fraternal sharing the fundamental resources—of water, of minerals, of energy, upon which our common life depends."

William Reilly notes that the trend among U.S. environmentalists toward global activity is markedly different from what it was in the sixties when nongovernmental organizations began working through the United Nations and other international agencies. Claudia de Moro Castro, executive secretary of Brazil's planning agency, remarked to Reilly not long ago that such groups were not in vogue "beyond the condition of U.S. support" for them. "They were American ideas seen as Yankee intervention," says Reilly. "Some were welcome, some not.

"In the 1980s, we are seeing local organizations growing up all over. I personally think that one of the most exciting features of international activity today is the degree to which nongovernmental organizations are helping their societies become less rigid, more decentralized, and more democratized." The observation is echoed by Harry Barnes, the U.S. ambassador to Chile, who predicts that future governmental leaders in Latin America will come from the environmental movement.

The Latin American phenomenon, adds Reilly, reflects a very different attitude toward environmental degradation. In the United States and Europe, "environmental problems are seen as an unpleasant, undesirable adverse consequence of economic activity. That is diametrically opposite from the way the developing world looks at environmental problems. The developing world sees them as consequences of no development at all or too little development. Degraded ecosystems, the failure to treat human waste, the failure to purify water, the prevalence of disease, malnutrition—all of these are associated with the lack of infrastructure, lack of capital investment, lack of sanitation services and hospitals.

"The language of poverty is the language of economics," says Reilly. "American environmentalists will make a profound mistake if they try to transfer the American experience to developing countries without recognizing that the fundamental imperative is development."

Environmental organizations such as the Sierra Club and NRDC have become more pragmatic in transferring the successes of the American experience abroad. Specifically, U.S. groups have moved away from international institutions in which the United States is one small voice, and generally outvoiced in the U.N. system, and reoriented themselves toward the U.S. government itself, urging it to be more environmentally responsible through its AID program and through its participation in the World Bank and multilateral development banks where the U.S. voice and vote is proportional to its funding. The Sierra Club even closed its New York U.N. office, which it had maintained for a decade.

McCloskey also notes that the psychology surrounding international issues has changed. "For years there was this long list of disaster scenarios filled with gloom and doom—so many of them that you just got depressed and didn't want to read about them. Then we got into computer models and feedback systems—remember the Club of Rome?—that nobody could understand. The message was that international issues are a scientific problem. We never did what as pragmatic reformers we did in this country."

Today's actions tend to be highly pragmatic and specific: Greenpeace lobbying the International Whaling Commission after having drawn worldwide attention to slaughter on the high seas; WWF funding a visitors' center in the hills outside Mexico City where monarch butterflies roost in the tropical forest—an overt swap of timber receipts for tourist dollars; NRDC and the international Pesticides Action Network pressuring the U.S. agricultural pesticides industry to set standards for the chemicals it exports; the Conservation Foundation assisting Costa Rican environmentalists to assess environmental problems and trends in their country.

But beyond these specific actions—and the most promising trend in today's environmental movement—is the emergence of a world view that expands the environmentalists' concern for nature to include the means of human survival. The implications of this new development have only begun to be explored and understood.

2

The
Monkeywrenchers

Dick Russell

THE SCENE: A remote wilderness area in the Siskiyou National Forest near Grants Pass, Oregon, May 1983. A construction company has begun bull-dozing an access road to about 150,000 acres of old-growth trees that the U.S. Forest Service plans to have clear-cut.

It is shortly before dawn when two burly, bearded men arrive and set up a log roadblock, then retreat temporarily into the dark forest. When sheriff's deputies, anticipating another round of confrontation with the group known as Earth First!, show up and remove the obstruction, Dave Foreman decides to use his body instead.

As a pickup truck grinds down the narrow dirt road, Foreman steps out and blocks the way. The truck slows but keeps coming, hitting him in the chest, knocking him back five feet. He rises to resist. This time the driver accelerates, backpedaling Foreman up a hill, faster and faster until he can no longer keep his balance. Falling, desperately he clings to the bumper, his legs trailing under the engine for over 100 yards until the pickup finally stops. As the workers pile out and surround him, shouting obscenities, the local police move in and place him under arrest.

Dave Foreman's knees suffered permanent damage, but four years later, Bald Mountain Road has yet to be completed. Three more blockades bought time for a lawsuit resulting in a temporary injunction.

The scene: The Faroe Island, a Danish territory in the North Atlantic, July 1986. For two weeks, a 200-foot converted trawler has been disrupting a traditional hunt, the yearly roundup of over 2,000 pilot whales by local residents who drive them into shallow bays where they spear and butcher them. Five crew members of the *Sea Shepherd II* have been arrested on shore, allegedly for customs violations. Only nine people, including a BBC documentary team, are still on board when an inflatable police boat pulls alongside, its officers staring up at barbed wire draped around the ship's rails to prevent boarding.

The *Sea Shepherd's* captain, Paul Watson, knowing that his ship is outside the island's three-mile territorial limit, foresees no problems. But as Watson steps out onto the starboard bridge, he sees a Danish officer standing fifteen feet below, pointing a shotgun at him. "What do you intend to do, shoot me?" Watson asks. The words are scarcely out of his mouth when a three-inch-long tear-gas shell whizzes within inches of his head.

Watson runs for cover into the wheel house as a barrage echoes against the steel bulkhead. He grabs a rocket flare, steps out again, and discharges it toward the police as the *Sea Shepherd II* heads for the open sea. For the next forty-five miles, a police gunboat and three inflatables pursue the ship through international waters, firing tear gas as the Sea Shepherds don gas masks and fight them off with fire hoses and distress flares.

As Butch Cassidy used to say to the Sundance Kid, gazing back across the hills at his constant pursuers, "Who *are* those guys?" Dave Foreman and Paul Watson are the founders of two groups at a new extreme of the environmental movement, whose dedicated cadres are willing to go out on many a limb or walk the plank in defense of forests and wild creatures. Earth First!ers sit in protest perches atop giant redwoods, and sometimes sink spikes into trees to make life difficult for the loggers and lumber millers; they lie down in front of bulldozers and other heavy equipment, and sometimes "decommission" them by dumping nefarious substances into the gas and oil. Sea Shepherds have claimed responsibility for sinking five whaling vessels of three nations, the latest a pair of Icelandic ships sent to the bottom of Reykjavik Harbor last November.

To adherents, these tactics are known as monkeywrenching and ecotage. To critics, they are a form of terrorism that gives all environmentalists a bad name. Greenpeace, which eschews the destruction of property, has compared the Sea Shepherds' Icelandic action with the French sinking of its *Rainbow Warrior.*

"Going out and pulling up a survey stake is not terrorism. It's simple self-defense," responds Dave Foreman. "The terrorists are the Forest Service with their plans to construct 75,000 miles of roads in existing wilderness areas over the next fifteen years, destroying elk and grizzly habitat. The terrorists are the ones cutting down thousand-year-old redwood trees to make picnic tables, and damming up wild free-flowing rivers."

Paul Watson speaks in the same vein of Icelandic, Norwegian, and Japanese ships violating a moratorium on commercial whaling, and of Japanese fishermen who, in their miles of drift nets, indiscriminately drown 1 million seabirds, 50,000 northern fur seals, and over 5,000 Dall's porpoises each year. "We consider ourselves the true conservatives," Watson says. "Our aggressive nonviolence is just doing what must be done, according to the dictates of our own conscience."

In style, these two men approaching middle age (Watson is thirty-six, Foreman, forty) tailor themselves well to their respective roles. Foreman is soft-spoken, but his fiery oratory leads Huey Johnson (founder of the Trust for Public Land and California's former environment chief) to call him "the most eloquent speaker in the environmental business." In his reddish beard, Levis, and cowboy boots, he bears no slight resemblance to a nineteenth-century mountain man. Watson, on the other hand, is a uniformed, well-pressed, modern-day John Paul Jones.

Both men came to their calling after departing established environmental organizations in frustration. Foreman resigned his position as a Washington, D.C., lobbyist for the Wilderness Society in 1980, fed up with what he considered political compromising. Watson, one of the founders of Greenpeace, left in 1977 after colleagues objected to his grabbing a seal hunter's club and throwing it into the sea, an act they regarded as violating their pacifist principles. The two have moved on to found what might be termed nonorganizations, even refusing regular salaries for themselves or any other full-time workers. Their criticisms of their more bureaucratic brethren have helped stir up a maelstrom within the movement that is seen as healthy by some, divisive by others.

David Brower, past executive director of the Sierra Club and chairman of Friends of the Earth, has spoken at Earth First! gatherings and provided them office space when needed at his new Earth Island Institute in San Francisco. Recently, Brower defended them against the terrorist label at a meeting of Montana forestry officials. "I said they're spiking trees which shouldn't be cut, by an industry that's only concerned about the short run," says Brower. "The environmental movement has gotten very drowsy, and I think Earth First! is giving it CPR. I admire people who put their bodies where their mouth is."

But Doug Scott, who once worked with Foreman at the Wilderness Society and is today the Sierra Club's conservation director, thinks that what his ex-colleague is doing now "may be a lot more fun, but Earth First!'s institution-bashing is rather destructive. I think it reflects impatience or lack of sense in understanding that in the world we live in, complex institutions are necessary."

Earth First! has attracted a legion of supporters over the past few years. Its formation seven years ago consisted of a half-dozen people camping under a full moon in the Mexican desert, dreaming of a wilderness preserve system and scheming "how to put a monkeywrench into the gears of the machinery destroying natural diversity" (as Foreman later wrote). Since then, it has grown to seventy-two chapters in twenty-four states, plus twenty-six individual contacts in nineteen more states. While still mainly on the West Coast (thirty-nine chapters, with twenty-two in California alone), Earth First! has spread across the United States and established liaisons in Mexico, Canada, Australia, Japan, England, and Scotland.

Earth First!—The Radical Environmental Journal, begun as a giveaway, eight-page newsletter, is today a twenty-six-page, $2-each organ of news, how-to's, and deep-ecology philosophy. Issued eight times a year, on the old Druid holidays, the paper claims about 10,000 avid subscribers. Foreman's *Ecodefense: A Field Guide to Monkeywrenching*, self-published in 1985, sold out its first edition of 5,000 copies, and was reissued last spring at $12 per copy, its contents doubled to over 300 pages through reader feedback.

The Sea Shepherd Conservation Society got off the ground with a $120,000 grant from Cleveland Amory's Fund for Animals and $50,000 from the Royal Society for Prevention of Cruelty to Animals (in Great Britain). Based in Watson's hometown of Vancouver, it has offices in Los Angeles, Washington, D.C., and Plymouth, England; publishes a quarterly newsletter; and boasts about 12,000 members. Though that is a far cry from Greenpeace's roster of 725,000 in the United States alone, it has captured the fancy of Hollywood. In 1983, when Watson was arrested during a blockade of a Canadian seal hunt, his $10,000 bail was posted by Mike Farrell, costar of the "M*A*S*H" TV series; and according to the Sea Shepherds, actor Jon Voight has expressed interest in portraying Watson in a film.

•

Driving to lunch recently in suburban Tucson, Arizona, where Foreman, his wife, and two other Earth First!ers maintain home base for what he terms "not an organization but a movement," Foreman described his flight from Washington, D.C., "where today's big conservation groups want to be decision makers instead of advocates for a particular point of view." He feels they have lost touch with the grass roots, deal with issues on an abstract level, pay their bosses excessive salaries, and have become much like the corporations and political bureaucracies they still think they are fighting.

Foreman, who was once an avid campus campaigner for Barry Goldwater, did not arrive at such conclusions overnight. Even in 1978, after seven years with the Wilderness Society, he would gaze out upon bustling Washington and consider, "Hey, if I play my cards right, I might be assistant secretary of the interior one day." But something happened the day the Forest Service issued its Roadless Area Review and Evaluation (RARE II). Foreman had spent months lobbying to protect more national forest land against logging, road building, and other developments, but of the 80 million undeveloped acres, RARE II recommended preserving only 15 million. "I realized then that our mildly reformist stance had blown it,"

Foreman recalls. "We'd been factual, rational, played the game, and got our ass kicked." When the Wilderness Society decided not to sue, fearing that even its minor gains might be lost, Foreman went home to the Southwest "to think it over."

For some time, friends in Washington's environmental organizations had been kicking around the notion of "a more militant group, that could be trotted out on occasion to take an extreme position and make them look more reasonable." In 1980, with the Reagan era looming, Foreman decided it was time.

Earth First!'s founders included two defectors from the Wilderness Society, one from Friends of the Earth, a disenchanted Forest Service employee who had been active in the Sierra Club, and a six-foot-six-inch ex-Yippie, Vietnam war protestor, and oilfield roughneck named Mike Roselle. Their inspiration was novelist Edward Abbey's *The Monkey Wrench Gang*, which described a band of environmental Robin Hoods out to halt the desecration of the southwestern desert by any means necessary. (Today, Abbey, who also lives in Tucson, is a staunch Earth First! supporter.)

"At first we were gonna try to work within the system," Foreman says, devouring a taco on the outskirts of Tucson. "But the more we got involved in Earth First!, the more we began to question the assumptions of technological civilization and to realize it isn't reformable. I think of us as being on the *Titanic*. There are icebergs ahead and nobody's in the pilot cabin; everybody's arguing about rearranging the deck chairs. I just want to make sure there are some lifeboats outfitted, you see?"

The Earth First! lifeboat is an attempt to "avoid the hierarchy and administrative bulge" that Foreman believes is weighing down most environmental organizations. In 1981, Earth First! began on a shoestring, as Foreman and a musician calling himself Johnny Sagebrush barnstormed cross-country through forty towns in two and a half months, making speeches, singing, and passing the hat. Today they operate on an annual budget of approximately $100,000, raised primarily through Foreman's publishing enterprise, which also sells T-shirts and cassette tapes. ("What would the antiwar and civil rights movements have been without music?" poses Foreman. "But where do you find it in any other environmental group?") There is also an Earth First! Foundation to manage tax-deductible contributions. Local chapters are completely autonomous; they fight their own battles, but are sent Earth First! publications for fund raising, get space in *Earth First!,* and receive periodic inspirational visits from Foreman or tactical visits from Mike Roselle's wandering "Nomadic Action Group."

Though Earth First! groups have sprung up in places like Los Angeles, where demonstrations against toxic chemicals are the primary thrust, the most likely place to find them is in the wild. Earth First! actions are both dramatic and headline-grabbing. Protestors have roosted atop giant Douglas firs slated for cutting in Oregon's Willamette National Forest. Outraged about elimination of grizzly habitat, they have dressed up in bear costumes and blockaded a Park Service bridge in Yellowstone. In Austin, Texas, six protestors were arrested recently for pitting themselves against a huge Forest Service bulldozer they dubbed "Godzilla."

Their civil disobedience often gets results. After the Texas media gave the bulldozer blockade wide coverage, the state's attorney general declared a moratorium on the clear-cutting, questioning the Forest Service's claim of "no significant impact" on the environment. In Siskiyou National Forest, where the local Sierra Club had lost its appeals trying to stop Bald Mountain Road, Earth First! revived the issue with a series of blockades. Eventually, forty-four people were arrested and the Oregon Natural Resources Council filed a successful lawsuit to stop the road.

"An Oregon wilderness bill has since released the largest, most diverse conifer forests in the West to more lumbering," says Foreman. "But we have created the old-growth issue and at least slowed down the juggernaut. The bill probably had a couple hundred more acres protected in it because of all the direct actions. What's going on is, the big lumber companies want to liquidate the old growth because it's easy profits for them. And the Forest Service is all for it, because even though they lose money putting roads in, logging off old growth, and replanting tree farms, that's their idea of putting it under management. But you know, a few of the loggers themselves have pioneered tree-spiking techniques—because they know that going in and just annihilating the old growth is eventually going to cost them their jobs."

To the timber industry, old growth is highly profitable. With tighter rings and fewer knots, it is the highest grade of redwood; it brings $1,500 to $1,800 per thousand board feet, nearly three times as much as the wood from younger trees. "The old-growth forest is dying anyway," shrugs a spokesman for Pacific Lumber. "It's better off cut and cleared away." The tree sitters and spikers do not see it this way, of course; as one says, "we're not going for scenic beauty but for biological diversity." In response to monkeywrenching, Louisiana Pacific offers rewards of up to $5,000 for information leading to the arrest and conviction of tree spikers.

The effect of Earth First!'s underground activities is difficult to measure. Foreman claims (here is where the nonorganization shines through) that as an organization, Earth First! does not advocate the monkeywrenching techniques detailed in a regular column of the newspaper and throughout *Ecodefense*. He personally claims to be opposed to guns, explosives, or major industrial sabotage. "It won't work because you're going head-on into somebody much more powerful than you," he says, "and I don't think you're ever justified in hurting innocent people. No, we're guerrilla fighters. We're recommending jiujitsu against the power structure." In this vein, *Ecodefense* instructs readers on how to halt vehicles, burn machinery, bedevil surveyors, and harass politicians, all under Foreman's tongue-in-cheek disclaimer: "No one involved in the production of this book . . . encourages anyone to do any of the stupid, illegal things contained herein."

The book's major contributors, according to Foreman, are a half-dozen individuals, most of whom he has never met. After shipping, Foreman burns all orders he receives for the book, "in case the FBI's going through my trash." In his "conservative" perspective, the cardinal rule of monkeywrenching is: don't get caught. His friend Howie Wolke, who served six months in jail for pulling up stakes, got overconfident and made a mistake, says Foreman; the second edition of *Ecodefense* contains a seventy-page section on "security precautions."

We are sitting in Foreman's living room now, surrounded by jammed bookshelves and artifacts from his travels. "Earth First! is successful because of something in the air," he continues. "A lot of people were ready for it. I see us as part of a natural process within the living organism of earth, combatting a disease. I just want to try to ensure that there are some pockets of natural diversity left after the self-destruction—because it's inevitable that this industrial bubble that's been constructed is going to burst over the next 100 years."

In another room, Earth First!'s only two full-time office staff are hard at work on the next issue of the paper. The headlines are pure Hearst:

"Exxon Invades the Northwoods," "War in the Wenatchee," "Yellowstone's Watergate," and "Occurrence in the Ferocious Isles—Sea Shepherd Takes on Whale Butchers." Reporters sport comic-strip bylines: Arthur Dogmeat, Doug Fir, and Bobcat. Columns dubbed "Tribal Lore" and "Nemesis News Net" appear alongside philosophical discourses by deep ecologists, bioregionalists, Greens, and Native Americans. The paper has become a forum for opinions from various schools of eighties environmentalism.

Past huge wilderness maps that line the hallway is Foreman's private office. Here, tucked neatly in the closet, animal costumes nestle among a pile of sleeping bags. He and his wife Nancy, a cardiac-care nurse whom he met when his "road show" went through Chico, California, a few years ago, try to get into the wild as often as possible. They were married in Yellowstone at Earth First!'s annual Round River Rendezvous in 1986.

Among Earth First!'s ranks is a group of teenagers Foreman calls the Little Wolves. "It was frustrating for a while," he says, "because everybody in Earth First! was in their thirties or older. We just couldn't seem to reach younger people. Now that's changed. I joke sometimes that a bunch of young women, all in their twenties, are taking over."

Every summer, those attending the Rendezvous plan an action. For example, nineteen people from nine states were arrested blockading the Yellowstone Fishing Bridge on behalf of the grizzlies. This past summer, after the Rendezvous gathering at Grand Canyon, twenty-one Earth First!ers were arrested while protesting a uranium-mining operation; that action made the "CBS Evening News." This, in Foreman's view, is the greatest value of Earth First!—the empowerment of people.

In 1985, when the Georgia Pacific Lumber Company began logging California redwoods in Mendocino's Sinkyone wilderness, Foreman dispatched Mike Roselle to rally some troops. "We're trying to redefine leadership," says Foreman, "to get away from the corporate mode, but also to avoid the bottomless pit of having no leadership with a so-called consensus process. A leader is someone who basically is able to sense the mood of a particular moment and articulate it better than anyone else. Mike is a real catalyst. He goes somewhere and something's going to happen. In the Sinkyone, there were a bunch of backwoods hippies who wanted to protect the area, but I'm afraid they'd never have gotten their act together without Mike."

Roselle spent a week training them in civil disobedience and soon had fifty people willing to risk arrest. Their battle came to center inside the seventy-five-acre Sally Bell Grove near the coast, around a 3,000-year-old-redwood. The Medicine Tree, oldest and largest of the region's redwoods, rests upon an ancient sacred site of the Sinkyone Indians, but Georgia Pacific was determined to log this valuable slab of old growth. The protestors' hope was to disrupt the operation long enough for a local environmental group to win a court injunction.

Over the first couple of days, they slowed things down considerably by dashing through the grove within range of falling trees. As one group of Earth First!ers was hauled away by police, another would take its place. At first, the tactic seemed to work; the injunction came through on schedule. But when the case switched courts in a change of venue, a new judge had to rule all over again. Georgia Pacific took this as a signal to resume, and the Medicine Tree was tops on its agenda.

That morning, after eight demonstrators wrapped themselves around the redwood's massive base and were dragged off by police, the authorities sealed off the tree with a barricade. As the loggers resumed cutting, Earth First! took to the surrounding woods, where the shrill blasts of their airhorns, like battle trumpets, competed against the roar of the chain saws. "We figured if they knew we were in there, they wouldn't drop the tree right on us," recalls Roselle. "But then some loggers felled a smaller redwood, which hit an oak that went down and pinned a woman who was with us—fractured her collarbone. They didn't show any sign that they'd let us stop 'em. So we came out and closed in again around the Medicine Tree. I got arrested for standing right in front of where it might fall. And then my friend Ron Mulvaney jumped over the barricade.

"I mean, they were getting real close to making their final cuts on that tree. They'd just removed the wedge when Ron went right in front of the chain saw. Another logger came at him with an axe, swung it and just missed his head. The cops moved in, picked Ron up, and started dragging him out of there. But the whole scene took about ten minutes and, as it turned out, bought us just enough time. Just as they got Ron out, the sheriff showed up and shouted through a bullhorn: 'Anyone who *continues* cutting this tree will be arrested.' The new judge had decided to retain the injunction. Ron's risking everything in those moments had saved the Medicine Tree."

The Earth First! brigade replaced the wedge in the tree to help keep it standing, and Native Americans from the region came to perform a healing ceremony. The subsequent legal action, which involved the Save the Redwoods League and the Sierra Club Legal Defense Fund, ended up establishing precedents for California forestry on the issues of cumulative impact and Native American sacred sites. Last December, the Trust for Public Land signed an agreement to purchase 7,100 acres of the Sinkyone wilderness from Georgia Pacific for addition to a state park.

•

The story of the Medicine Tree would seem familiar to Paul Watson and the Sea Shepherds. Indeed, says Watson, his most crucial formative experiences were running away to sea when he was young, and going to Wounded Knee, South Dakota, in 1973. The American Indian Movement (AIM) was then occupying the Sioux reservation for seventy-one days in protest of government treatment, and two of its members had been killed by U.S. authorities. Watson volunteered as a medic. It was there, he says, that he "learned to conquer the fear of dying." When the occupation ended, Watson received honorary initiation into the tribe. Guided by two medicine men, including the grandson of the legendary Black Elk, he experienced a vision that was to change his life.

"I suddenly saw myself in a grassy, rolling field, gazing into the eyes of a wolf," Watson remembers, sitting in his captain's uniform outside the Los Angeles Press Club, where last December he announced a new campaign to confront the Japanese drift-net fishermen. "The wolf looked at me, then into a pond, and walked away. When I told the Sioux what had happened, they gave me my Indian name: Gray Wolf/Clear Water. Then I went back into the vision, and saw a buffalo standing on a ridge. It began to speak to me. And as it told me that I must protect the buffalo of the sea, an arrow came and struck it in the back. Attached to the arrow was a cord, symbolic of a harpoon."

Not knowing what it all meant, Watson went home to Vancouver. Six months later, a fellow he had known when they organized protests against nuclear testing came to see him. Robert Hunter said that a few friends wanted to turn their old "Don't Make a Wave Committee" into a new organization whose mission would be to save whales. Soon Watson was among the leaders of Greenpeace, up against the Russian harpoons from a little inflatable Zodiac that would capture the imagination of the world.

While Watson was at Wounded Knee, another future Sea Shepherd leader, Benjamin White, was living in Nevada with the renowned medicine man, Rolling Thunder. "He told me, in interpreting a couple of dreams I had, that my totem animal was a coyote," White remembers. "More and more, I came to see my purpose here as being defense of nature. There was also the way the Indian people faced confrontation. I was the only white man inside the Bureau of Indian Affairs when AIM took it over for three days in 1972. These people were actually willing to die, right there, for what they believed in. It was so different from what I'd seen at so many Vietnam protests. I'd never encountered the dedication I saw in the native people, and I think that really is part of the power of Sea Shepherds." White, now thirty-five, read a magazine article about the Sea Shepherds "and figured somebody talked my language." He joined Watson in 1981 en route to the Siberian coast, and is today the group's Atlantic director.

The Shepherds speak with feeling of their close encounters with whales and dolphins. Watson's eyes blaze as he talks of swimming among dolphins. "They can literally see inside you with X-ray vision; you can feel it penetrating your body. And whales give off almost a vibration that they mean you no harm. To touch them without fear is one of the most incredible experiences anyone could ever have." Adds White: "To look into a whale's eyes, and know it is regarding you intelligently, changes your life. To me, whales are like the canary in the mine; if they go, that's it for all of us, the whole human race."

No matter how irrational some may find the Sea Shepherds, the leadership is intent on staying grounded in practical realities. All support themselves with outside jobs. White, a husband and father, is a tree-care specialist whose ten employees keep the business going while he is away at sea; he combines his arborist and Sea Shepherd offices in Fairfax, Virginia. Scott Trimingham, the Sea Shepherds' president, is a toy buyer in Redondo Beach, California. Watson, who earns $1,200 per speaking engagement, plows his earnings back into the organization. Crew members signing on to *Sea Shepherd* expeditions must contribute as much as $1,500 for the privilege of putting themselves on the line; about 50 percent of the ship's expenses are financed by the crews. Since 1980, over 350 different people have taken part in campaigns. All of this, says White, "keeps us honest; it's good to stay hungry."

Like Foreman, Watson is adamant about "avoiding stifling bureaucracy. We don't have meetings; we have get-togethers maybe once every few months where we talk things over." This spring, the Sea Shepherds finally initiated a direct mail fund-raising effort, but still have no set membership fee. "You send in a dollar or $1,000, you become a member in equal standing," says Watson. "Our membership grows slowly by word of mouth." Although the group's annual budget now runs between $175,000 and $200,000, Watson adds, "We've never really had more than $3,000 in the bank at any one time." Their most vehement supporters are apparently women over 65. The Faroe Island campaign was funded mainly by English schoolchildren, who raised about $25,000 in a walkathon.

What appeals to the old and the young (and some in-between) is the very unorthodoxy of the Sea Shepherds. Somehow, it often works. "We've always relied on our intuition," says Watson. "We find the seal hunts based on where we feel the ships will be. Any time I've ever made a decision against my gut feelings, it's been a disaster."

There are a number of remarkable tales about Watson, an anti-Ahab of almost mythic proportion. For example, in 1981, a Soviet destroyer was pursuing the *Sea Shepherd II* through Siberian waters, still three hours from U.S. territory. The Shepherds were there to document illegal Soviet killing of endangered gray whales for use as feed in commercial fur farms, and the ship was racing home with the film. The destroyer, which was more than twice as fast, kept cutting across the *Shepherd's* bow, and soon was joined by a helicopter that began dropping flares. As the Soviets started the engine on a launch suspended halfway up the destroyer, Watson ordered the crew to grease the gunwales under the barbed wire that ringed the ship's deck, hoping further to dissuade a boarding party.

"The captain of their ship radioed over," recalls White, "and said '*Sea Shepherd II,* stop your engines and prepare for boarding by the U.S.S.R.'

Paul radioed them back, 'Stop killing whales!' The Russians sent their message again, and again Paul told them, 'Stop killing whales!' Suddenly all hell broke loose. I was standing on the foredeck. I looked over at our crew on the bridge, which was listening to Paul over the radio, and they were cheering like crazy. Then I noticed the crew on the stern were cheering wildly, too. Except they couldn't have known anything about the exchange with the Russians. They had seen something else. Out of the blue, a California gray whale had come up to the surface, right between our two ships." White continues: "Everyone took this as an omen that we would be okay. And within a few minutes, the destroyer backed off and just followed us out of Soviet waters."

An equally strange sequence of events occurred in July 1979, when the *Sea Shepherd I* set out to hunt down the *Sierra*, a whaler that operated outside the laws of the International Whaling Commission (IWC) by disguising its ownership and bribing Portuguese and Spanish authorities. The ship, under contract to the Japanese to deliver an annual 3,000 tons of frozen whale meat for a gourmet food market, regularly wiped out whole herds along the southern Africa coast. Watson had only a vague idea where to find it.

Unknown to Watson, just as he left port in Europe, a friend from Wounded Knee returned to the Sioux reservation in South Dakota. Meeting with Leonard Crow Dog, a medicine man, David Garrick explained what Watson was doing and asked for guidance. Crow Dog answered: "Gray Wolf will find the enemy, and be led to the enemy by the turtle."

"Our ship was a day out of the Azores," Watson remembers. "I later learned it was this same afternoon when I stopped for four hours to let thousands of migrating loggerhead turtles go by. If this hadn't happened, I would never have intercepted the pass of the *Sierra* the next day, 200 miles out to sea. I'd have already sailed right past them."

The *Sea Shepherd I* pursued the *Sierra* back towards Lisbon, where Watson informed the crew that he intended to ram it. When he gave the choice whether to remain on board, all but two of the nineteen-man crew decided to leave. The attack sent the whaler into port for major repairs. A Portuguese destroyer pursued the *Sea Shepherd I,* and Watson, "being then somewhat naive and feeling I could win in court," allowed the vessel to be confiscated. A few months later, he learned from the American

Consulate that the Portuguese intended to turn the *Sea Shepherd* over to the whaling company. After an unsuccessful attempt to steal it back, Watson and a partner sneaked aboard, opened the seacocks, and scuttled their own ship.

Shortly thereafter, just as the *Sierra* was reoutfitted and ready to resume whaling, it too went to the bottom of the harbor. A three-man demolition team had attached a magnetic mine and blown a ten-foot hole in the hull, sinking it in ten minutes. "A crew of individuals did it, paid for by another individual," says Watson. "Our connection was to fill them in on details. I only met with one person who was involved, and to this day I don't know who the others were." Two Spanish vessels that had been taking fin whales beyond quota limits met a similar demise during this same period. Warner Brothers advanced Watson money toward a film (a project that never got off the ground) that enabled the Sea Shepherds to purchase another boat.

Besides the near-miss in the Faroe Islands in 1986, Watson's closest brush with death followed his disruption of a Norwegian/Canadian harp seal hunt by spraying green dye on the animals, destroying the value of the pelts. Sealers captured and hogtied him, dragging him across the ice and threatening to toss him into the sea. Watson shrugs off the dangers of his chosen profession.

"Putting my life on the line is not really a big deal when you consider that 100 million people in this century have died in wars over real estate," he says, "and they've defended that action in the name of some abstract thing called patriotism. So I don't think it's unrealistic to expect people to risk themselves to protect a species of animals that took 100 million years to evolve. In my opinion, it's a much more noble cause."

In recent years, the Sea Shepherds have scored several major victories. In 1982, they negotiated a halt to the slaughter of dolphins off Japan's Iki Island. The Irish government shut down a rifle hunt of gray seals when some British Shepherds camped among the herds and placed themselves in the path of the bullets. In 1983, the last large-scale Canadian and Norwegian seal hunt was stopped when the Sea Shepherds blockaded the gulf (the International Fund for Animal Welfare conducted air reconnaissance to help locate the sealers), then worked with other groups to coordinate a European boycott of the pelts. In 1984, they started Project

Wolf to fight the killing of wolves in British Columbia, and bought two islands in the Orkney archipelago off Scotland for seal sanctuaries. Now, working with the California Department of Fish and Game, they have started a rescue team to free migrating gray whales that get entangled in fishermen's gill nets. Last May, in a major turnabout prodded by Greenpeace and Sea Shepherd encounters, the Soviet Union announced that it was halting all commercial whaling.

In Washington, Ben White meets weekly with the Monitor Consortium, a coalition of animal rights and environmental groups. But, while he believes legislative avenues should be pursued, he does not place exclusive faith in them. Since January 1986, the IWC has had a four-year moratorium on all commercial whaling, permitting only a limited take for scientific research and native peoples' subsistence. Iceland, which has continued to sell meat from the 120 "research" whales it takes each year, was threatened with U.S. sanctions last July unless it ceased. It did not, and the United States did not follow through. "So what were we supposed to do?" asks Peter Wallerstein, Sea Shepherd Pacific director. "The environmental groups that pushed for this moratorium for fifteen years are sitting back and not even educating the public about its being violated."

Last fall, Rodney Coronado came to Watson with a direct-action proposal. Coronado, twenty, had admired the group since he was twelve, and joined the crew immediately upon graduation from a northern California high school. Watson gave him the go-ahead to make plans so long as he adhered to these guidelines: no weapons or explosives, no action that could cause personal injury, no resisting arrest if apprehended, and be prepared to take responsibility for your act.

Coronado and a friend arrived in Reykjavik last October, where they were hired by the whale meat packaging factory; hidden in his belongings, Coronado had a dog-eared copy of *Ecodefense*. On November 10, a page-one story in *The New York Times* revealed that environmental militants had sneaked into the engine room of two of Iceland's four whaling ships, opened the valves, and sent sea water pouring into the holds. Both ships were at the bottom of the harbor. It was soon revealed that the saboteurs also had wrecked Iceland's plant for processing whale oil and other by-products.

The Icelandic incident brought the Sea Shepherds international attention. Coronado had flown to Europe before the sinking was discovered, and Icelandic authorities were demanding his and Watson's extradition. According to Watson, several wildlife groups privately expressed their support, but publicly condemned the act. The International Fund for Animal Welfare's chief executive called to congratulate Watson, though the same organization's British office was castigating him as "a threat to the entire movement."

•

But it was Greenpeace, whose Vancouver office called the raid an act of environmental terrorism, that often appeared the most intent on distancing itself from the Sea Shepherds. Besides a long-standing rivalry with Watson, there was the memory of the tragic event that shook Greenpeace to its core: on July 10, 1985, two bombs had sent its *Rainbow Warrior* to the bottom of a New Zealand harbor, killing a photographer on board. The French Secret Service later was found responsible, having retaliated against Greenpeace's peaceful protests of nuclear testing in the South Pacific.

"This struck right at the heart of Greenpeace," said Peter Bahouth, chairman of the U.S. board, who was in New Zealand at the time. "We weren't prepared for the emotions involved or the politics, and we've been scrambling ever since. Whether we like it or not, we're a changed organization." Since then, Greenpeace has been pulled in two directions — toward international affairs, and accommodation with bigness and bureaucracy; and toward a closer relationship with the grass-roots groups that its own history helped to inspire.

Greenpeace is also experiencing a surge in membership and funding. Drawing more than 10,000 U.S. members a month, it now stands at about 725,000 and its 1987 budget tops $16 million. [The international membership is in excess of 2.5 million with a total budget of about $35 million.]

But along with increased visibility and higher expectations has come a rapid turnover in the U.S. branch. Of the twelve top staff positions, ten have changed hands in the past year and a half. "The positive aspect is, people don't have real strong bastions to defend," says David Chatfield, who departed Friends of the Earth to head Greenpeace's San Francisco office a year and a half ago, on what he calls "the unification ticket."

"Two powerful elements needed to come together," Chatfield explained recently. "One was the success factor—having become a major national organization. The other was that much of the support and energy had always come from highly localized sources—the ability to move quickly and take risks. But the huge membership list was completely divorced from the regional effort, and vice versa." The result has been greater centralizing of administration in Washington, by dissolving the separate regional corporations, and a program planning process that now starts from the bottom up. While Greenpeace-USA's executive director, Steve Sawyer, admits he worries "about catching the Potomac fever and developing tunnel vision," he says that the days of political infighting are over "and we have more money to spend on programs because we're not constantly reinventing the wheel."

It is one of the ironies of eighties environmentalism that Greenpeace is acquiring a mainstream image by comparison with groups like Earth First! and the Sea Shepherds. While these favor working with a few people on local actions as far away from the so-called power center as possible, Greenpeace is more entrenched in Washington and more involved than ever in such complex international issues as ocean incineration and Antarctica. While the Shepherds announce dramatic plans to battle Japanese drift-netters on the high seas, Greenpeace fights against the same fishermen by suing the National Oceanic and Atmospheric Administration over lax restrictions on the porpoise take.

"We probably have more actions than ever before in Greenpeace around the world. The difference is they're very much connected to an international strategy," says Peter Bahouth. "We're talking now about World Bank funding, biocides, acid rain as part of a bigger air pollution problem, huge ecosystems that encompass the Great Lakes or Mediterranean, and disposal attitudes. We've gone from taking on individual pipes and chemicals to realizing that you can't really win the battle that way. You've got to make a much broader approach, attack whole classes of chemicals. The question is always, how effective is what you do on a large scale? We need to continue doing actions, but we've got to keep making quantum leaps in our way of thinking or we'll stagnate."

On a smaller scale, however, Greenpeace is faulted by people like Michael Belliveau, coordinator of California's grass-roots toxics groups,

Citizens for a Better Environment. "I like the Greenpeace folks, but they've had a certain inconsistency," says Belliveau. "One of the traps that Greenpeace has fallen into in the past is to start viewing various types of action as the ends rather than the means. I know several examples where there was no real planning or base-building in the community. They'd go in and do an action, no follow-up, just left town." Aware of such criticism, Greenpeace has inaugurated a computerized toxics bulletin board for grass-roots groups around the country.

In recent months, Greenpeace also has been trying to come to terms with the deep ecology philosophy that underlies Earth First! and Sea Shepherds. Sawyer is adamant that Greenpeace is not involved in any quest to define a philosophy, saying, "We've always avoided that like the plague. David McTaggart [Greenpeace's international chairman] and I are firm believers that as soon as you write, codify, and make something into a manifesto, it dies." Yet, a recent Greenpeace leaflet read: "Humankind is not the center of life on the planet. Ecology has taught us that the whole Earth is part of our 'body.' " An East Coast consultant, Ralph Whitehead, urged Greenpeace leaders "to go talk to New Age people and shape a message that fits that agenda." Soon after, the *Greenpeace Examiner* featured an article by Green philosopher Fritjof Capra, suggesting that "an outdated world view is responsible for the global crisis we face" and that Greenpeace embodies an alternative. For about a year, says one insider, Greenpeace's disarmament staff "sat around and discussed that idea."

At the same time, Greenpeace has established a relationship with Earth First! by utilizing Foreman's right-hand man, Mike Roselle, as one of its own direct-action planners. Roselle readily admits his own debt to Greenpeace, the onetime slingshot band which has become an environmental Goliath in fifteen short years. Recalling a 1986 Greenpeace action where he and a crew invaded an underground nuclear test site in Nevada, Roselle says, "When the military sent down these guys in helicopters to nab us, we heard they had orders to shoot to kill. I was standing there with a machine gun pointed right at me, totally unafraid, because I knew they knew I was with Greenpeace. That's reputation, see? If I'd said Earth First!, I'd probably have been dead meat."

Things have not been as smooth between Greenpeace and the Sea Shepherds, where spokespeople for the two groups sometimes have taken to denouncing each other's methods. Overall, however, the interconnections between direct-action groups seem to be increasing. Watson's answering machine in Vancouver lists Earth First! along with the Shepherds, and Foreman says that some Earth First!ers soon may be journeying on Sea Shepherd missions. Last year, Greenpeace canvassers trudged door to door carrying Earth First!'s petition to save the mountain lions.

In what Foreman considers the best example of pooled resources, Greenpeace, Earth First!, and the new Rainforest Action Network (RAN) teamed up last fall for a demonstration at World Bank headquarters in Washington. The occasion was the bankers' annual convention. Greenpeace designed and paid for a huge banner, emblazoned with a growling jungle cat, that read: "World Bank Destroys Tropical Rainforests." Mike Roselle, one of the founders of RAN (now operating out of Brower's Earth Island Institute), traveled East to coordinate the logistics. Under his tutelage, RAN activists climbed the fire escape of an adjoining building, raced across the rooftop, rappelled down the World Bank, and unveiled the banner just as the bankers were pulling up in their limousines. Three Greenpeace members who participated were also arrested.

With a growing acceptance of their different roles, might these organizations recognize that they have something to learn from each other's experiences? Greenpeace's Chatfield had a recent warning to offer Earth First!. "Being anarchistic and priding themselves on that," he said, "if they become very successful, they're gonna have a real problem to deal with. The more resources anybody has, the more conflict there is over how to use them and who chooses what."

Already, Roselle has found himself in conflict with the Earth First! Foundation over his roving Nomadic Action Group. "We needed climbing gear, banners, transportation equipment to get into remote areas," says Roselle, but the foundation insisted it could only fund education and research.

Roselle says they reached an accommodation at a meeting in April. "We've agreed that the two factions in Earth First! desperately need each other," he says. "Basically my group will be a totally separate thing. I'll raise money, too, but it won't be tax-deductible, but strictly a grass-roots financial support network. The foundation's money will go into projects like a legal task force, which is budgeted for the first time. I'm comfortable with the balance, but it's not been easy to achieve."

While Roselle frets over the foundation's centralizing tendencies, some local groups in northern California have set up their own fund-raising apparatus. At the same time, Earth First! chapters in Austin, New York, and Seattle have started doing more organizing and research around the rainforest issue. But as Earth First! grows, spreads and, as Greenpeace has, takes on more issues, Roselle is convinced that the group's roots will not be allowed to wither. In fact, he says, monkeywrenching is quietly on the rise.

"Just heard yesterday," he said gleefully over the phone recently, "that there's a lot of spiking of old-growth Ponderosa on the Kaibab Plateau in Oregon, which has caused serious problems at the mill, though nobody knows to what extent. On BLM [Bureau of Land Management] land, the powers-that-be have been felling trees and just leaving 'em there, to try and show us they won't stop. But it is definitely getting harder for them to market. There's an increasing war of nerves."

In July, the Sea Shepherds took off in pursuit of the Japanese drift-netters, carrying a twenty-seven-member crew in a new 180-foot vessel christened the *Divine Wind*. Slowed down first by a burst pipe in the engine room, then by severe storms in the Aleutian Islands, Watson received some astonishing news. The Japanese, already restricted by a Sierra Club lawsuit from fishing in U.S. waters for twenty days during the height of the tuna season, apparently decided to avoid a confrontation with the Sea Shepherds in international waters; dropping off their observers in the Aleutians, the fleet returned to Japan without putting out its nets again. Next stop for the Shepherds is Antarctica this winter, where they intend to go up against Japanese "research" whalers authorized to kill 825 minke and 50 sperm whales.

•

Skeptics say that such actions are no more than symbolic gestures. Congress just gave the Forest Service $229 million more for wilderness road building over the next three years; Iceland just raised the two sunken whalers and has plans to refurbish them. But Watson points out that for days after the Icelandic episode, everyone in the Canadian pubs was talking not about ice hockey, but about whales. The Sea Shepherds and Earth First! see their role also as raising consciousness. And while many environmentalists will continue to disavow their means, some have found themselves undergoing a shift in thinking.

"As an observer, I think they're fantastic," says Alan Davis, president of the Conservatree Paper Company and the California League of Conservation Voters. "We're at a point where there's a complacency in terms of our constituency, not the kind of apparent fervor that's existed in the movement before." Adds Phillip Berry, a prominent Oakland lawyer and twice president of the Sierra Club: "Well, I wish they wouldn't do it, because I think it makes us look a little fringey. But frankly, there is some point at which civil disobedience might come out in all of us. What if Reagan pushed through a proposal to build a dam in Yosemite Valley? Would I go and stand in front of a bulldozer or put sugar in their gas tank? Yeah, I think I would."

Last March, at an environmental law conference in Oregon where Brower and Foreman were the keynote speakers, noted law professor Charles Wilkinson ended his remarks by saying that driving deals is needed—but so is driving spikes. Brock Evans, a vice president of the National Audubon Society, recalled some old-growth battles he had been involved with in the Pacific Northwest and concluded that he was so frustrated about the national forests, he soon might go before the bulldozers himself.

To Foreman, such statements are a signal that the message is getting through. "The planet is either a collection of resources, or it's alive," he says. "That's the real choice. I'm happy there's something bigger than me out there: the natural world. And I'm not gonna throw my life away, but if my death would save the last of the grizzly bears, then fine."

3

The Legal Eagles

Tom Turner

WASHINGTON, D.C., IN THE EARLY 1960s was in ferment. Young Jack Kennedy had been elected president and appointed an eloquent Westerner, Stewart Udall, to run the Department of the Interior. Big Ed Muskie of Maine, chairman of the Senate Public Works Committee, was the congressional Democrat most concerned with ever-worsening problems of pollution. His Republican counterpart was John Sherman Cooper of Kentucky. They were determined to stop American industry from ruining the public health and poisoning land, air, and water, but they had little to work with besides their powers of persuasion.

Accordingly, they evolved a jawboning process called the conference. Captains of the major polluting industries would be called in to discuss ways of reducing pollution from manufacturing processes and products. The intended outcome of the conferences was an agreement on what pollution-control measures were economically and technically feasible.

These discussions produced little in the way of concrete improvements and convinced most lawmakers that the country would never achieve any significant degree of pollution control without standards that required companies to go beyond economically and technically feasible control technologies toward the cleanest technology available. The decision to "drive" technology began with a system of permits issued under the Water Quality Improvement Act.

By the end of the decade, public concern over the looming environmental crises was swelling fast. Earth Day was just around the corner.

It was against this background that the country's tiny environmental law establishment met at a conference center called Airlie House in the foothills of Virginia's Shenandoah Mountains. It was autumn 1969. Fifty people attended, some fresh out of law school and most others early practitioners of a field not even referred to as *environmental law.* The Environmental Defense Fund was by then in business, staffed wholly by volunteers, and was starting to use the courts to make life miserable for the pesticide industry. The Natural Resources Defense Council and the Sierra Club Legal Defense Fund were as yet little more than gleams in the eyes of a handful of idealistic young attorneys.

The conferees were full of fire and enthusiasm. They talked passionately about various legal approaches that might be tried to defend the environment. Some advocated relying on the doctrine of public trust. A feature of English common law on which much of our legal system is based, the doctrine holds that the governing body has certain responsibilities to care for and protect those things held in common by all citizens. This might be extended to lakes, rivers, parks, marshes, wildlife, and so forth.

Many suggested building on common law, which draws on previous decisions and opinions of the courts, rather than statutes, to bolster a claim of injury. Consider a factory that discharges noxious chemicals into a creek. It is well established in common law that a person who injures another can be sued for compensation. Under common law, then, a property owner downstream from the factory could go to court seeking compensation for damage to his property and his health caused by the polluted water. Wholesale reliance on common law presumably would not have required the network of laws and agencies that now exists. The courts, on the other hand, would have been called upon to decide if avoidable injury had been inflicted by another party. If the plaintiff was victorious, his or her case could be cited and built upon by others claiming similar sorts of injury. At the time, many people believed that such an approach to environmental protection would have clogged the courts with thousands of cases and resulted in hundreds of different interpretations of similar cases. The feeling among most conferees was that common law would not provide sufficient national perspective or be comprehensive enough to improve environmental quality.

Some delegates suggested that the lawyers turn their attention to establishing a constitutional right to a healthy environment by bringing cases based on the Fourteenth Amendment to the Constitution. That amendment forbids infringement of liberty or taking of property without due process of law, the general idea being that damage inflicted on the environment by loggers, miners, polluters, or others could be headed off by suits that would argue due process was being abused.

Others, notable among them the former Vermont Governor Philip Hoff and California Congressman Pete McCloskey, anticipated that the public's growing concerns about environmental quality would lead to demands for government intervention. "We are going to be deluged with environmental legislation from all quarters in the period of the next few years," Hoff told the conferees. "Conservation is going to have a very, very sharp focus in Congress in the next few years," agreed McCloskey. How right they were.

Within a few months, President Nixon had signed into law the National Environmental Policy Act (NEPA). A torrent of federal laws followed, among them the Clean Air Act, the Clean Water Act, and the Endangered Species Act. The lawyers suddenly had a whole new arsenal of weapons. New initiatives in public trust, common law, and constitutional protections were deferred, as the lawyers delved into the statutes. And from that period on, litigation would become one of the most important tools conservationists had at their disposal.

But new laws would not be of much use without the right to bring a case to court in the first place, a right that had been denied citizens, in many instances, until the mid-sixties.

Until then, lawyers had had a very difficult time persuading courts to hear cases involving land-use disagreements, for example, since the plaintiffs could not prove—indeed, they did not claim—that a certain decision would harm them economically, the traditional requirement for what is known as "standing to sue."

Then, in the early sixties, the Consolidated Edison Company proposed to build a huge pumped-storage plant on the Hudson River at Storm King Mountain near the U.S. Military Academy at West Point. The project became the premier environmental crusade in the state, but the Federal Power Commission ignored the protests and gave the project its blessing.

An organization called the Scenic Hudson Preservation Conference and three towns in the area then filed suit against the FPC. As expected, the commission asked the court to dismiss the case since the plaintiffs claimed no pecuniary interest in the project. But in a landmark decision, the Court of Appeals for the Second Circuit ruled that Scenic Hudson should be allowed to bring suit owing to its "aesthetic, conservational, and recreational" interest in the area. The idea of a noneconomic interest being sufficient to establish standing was born.

The Scenic Hudson case also broke new ground by requiring that agencies consider the alternatives (including the alternative of doing nothing) and validating the conservationists' claim that scenic values should be weighed against economic values. Soon after, in a Sierra Club case to block a ski development at Mineral King Valley in the Sierra Nevada, the Supreme Court ruled that all the club had to do to establish standing was to allege that the interests of its members would be harmed. With some exceptions, standing has not been a serious hurdle since.

NEPA ushered in a new era of environmental awareness by requiring federal agencies to include environmental protection in all their plans and activities. And it invented the environmental impact statement, the EIS, a methodology and instrument for assessing the likely effect of projects agencies intend to build, finance, or permit. Most important, the EIS required agencies to investigate alternative ways of accomplishing the same purpose. (The alternative to building a power plant along a scenic coastline, for example, might be a program of energy conservation that would eliminate the necessity for a new facility.) NEPA also created a

Council on Environmental Quality to advise the president. The center of considerable influence throughout the seventies, CEQ has withered under the Reagan administration and awaits rejuvenation.

To complete the burst of activity, Richard Nixon created the Environmental Protection Agency, which has since become the biggest bureaucracy in government outside the military. When Congress passes a pollution-control law, it is most often EPA which first must issue regulations and then enforce the law. Failure to do so can bring swift legal action, but the action may be against the government itself (as it often has been) for failing to issue regulations on time, or failing to set a proper standard, or failing to enforce its own regulations once they are in place.

Not surprisingly, attorneys, who had played supporting roles in environmental struggles up to that time, took center stage. According to some, it changed everything. Rick Sutherland, cofounder of the Los Angeles-based Center for Law in the Public Interest and for the past ten years executive director of the Sierra Club Legal Defense Fund, says simply, "Litigation is the most important thing the environmental movement has done over the past fifteen years."

David Sive, who was involved in the famous Scenic Hudson case and is one of the field's pioneers, is more encompassing: "In no other political and social movement has litigation played so important and dominant a role. Not even close."

John Adams, the first and so far only executive director of the Natural Resources Defense Council, echoes the sentiment, saying that "the legal victories won in the late sixties and early seventies formed the foundation on which the modern environmental movement is built."

And James Moorman, the first executive director of the Sierra Club Legal Defense Fund now in private practice in Washington, agrees, adding that litigation "made the bipolar system of regulator and regulated tripolar: regulator, regulated, and the public."

Step outside the environmental law establishment, however, and you will find differing opinions. Ed Frost, for example, who served five years as general counsel to the Chemical Manufacturers Association, thinks environmental litigators have engaged in substantial overkill, particularly with respect to deadline suits. These are lawsuits, usually filed against the Environmental Protection Agency, for failing to meet deadlines set by Congress for promulating regulations for any one of a host of pollutants. Environmental attorneys find them attractive, because they are generally easy to win and help to focus public attention on problems they may have thought were being handled. "The deadline suits have chastened the agencies," he said, "but they haven't really helped set priorities or clean up the environment. They've left things in chaos. Compliance has been random." The agencies, he maintains, simply cannot handle all the work assigned them by Congress.

Norman Bernstein, associate counsel of the Ford Motor Company, maintains that at least some of the litigation over toxic waste cleanups has resulted in excessive transaction costs, such as legal fees and engineering studies, that contribute little to a clean environment. He suggests that the top-down style of environmental management is beginning to impinge on basic constitutional protections afforded both individuals and companies; for example, the right to do what you want with your own property. Though courts generally have upheld environmental laws on constitutional grounds, such corporate complaints are widespread.

The enactment of NEPA in 1970 sparked what Tony Roisman, who tried many of the early landmark lawsuits as a staff attorney with NRDC and in private practice, calls the "Golden Age of environmental law." "Government couldn't write a passable EIS. You could stop almost anything. Injunctions flowed like water from the courts."

One of the early injunctions came in a case brought by Roisman and others, on behalf of the Calvert Cliffs Coordinating Committee. Calvert Cliffs was a 1,000-megawatt nuclear power plant the Baltimore Gas and Electric Company wanted to build at Lusby, Maryland, overlooking Chesapeake Bay. The Coordinating Committee sued the Atomic Energy Commission for violating NEPA when it issued Baltimore Electric a construction permit.

Following the passage of NEPA, the Atomic Energy Commission had produced a set of rules to guide the agency's compliance with the new law. Roisman argued that the rules took NEPA so lightly that the AEC allowed itself to proceed as usual, ignoring environmental values whenever it felt like it. Further, he argued that the environmental impact statement on the power plant was deficient in that, among other things, it failed to consider the environmental impact of discharging heated water into Chesapeake Bay. What may have seemed a desperate legal ploy to a layperson proved central to the outcome of the case. The Coordinating Committee argued that NEPA was more than a procedural checklist; that it required a broader, holistic, ecological approach to development. Thus, the power plant's off-site impacts, on the bay itself and the aquatic life it supports, were as important as its on-site impacts. If upheld, *Calvert Cliffs* would be a legal quantum jump above *Storm King,* which first established the requirement of interdisciplinary analysis.

The AEC, not unexpectedly, took the straight and narrow view that NEPA was an internal protocol and that Congress never intended it to be used as a means of overturning agencies' decisions. The Court of Appeals in Washington, D.C., however, sided with holism, and the project was halted.

In his celebrated opinion, Judge Skelly Wright gave no quarter to AEC's defense:

> We must stress as forcefully as possible that this language does not provide an escape hatch for footdragging agencies; it does not make NEPA's procedural requirements somehow "discretionary." Congress did not intend the Act to be such a paper tiger. Indeed, the requirement of environmental consideration "to the fullest extent possible" sets a high standard for agencies, a standard which must be rigorously enforced by the reviewing court.

A scant six years after the euphoria of *Calvert Cliffs,* however, an up-and-coming Supreme Court justice wrote a decision that knocked the NEPA express off the rails. NRDC had sued the Nuclear Regulatory Commission claiming that the agency should have included reprocessing and waste disposal in the environmental impact statement for the Vermont Yankee nuclear power plant. The Court of Appeals had agreed. But Justice Rehnquist said the court had overstepped itself:

> Nuclear power may someday be a cheap, safe source of power or it may not. But Congress has made a choice to at least try nuclear energy, establishing a responsible review process to which courts are to dis-

play only a limited role. The fundamental political questions appropriately resolved in Congress and in the state legislatures are not subject to reexamination in the federal courts in the guise of judicial review of agency actions.

In other words, NEPA can stall projects until adequate studies are done, but once the studies are complete no court can force government to choose one alternative over another, even if one is clearly superior environmentally. That discretion—in the court's view—properly belongs to agencies or the legislative branch.

It was devastating. Roger Beers, a one-time NRDC lawyer now in private environmental practice in San Francisco, recalls, "I had a suit against the Atomic Energy Commission at the time, arguing that a nuclear plant shouldn't get a license until the AEC could say what it would do with the [radioactive] waste. It was a very good case so far as common sense was concerned. The argument went very well. Then *Vermont Yankee* came down and it sent a signal that courts were not to meddle with agency decisions. We lost."

Nevertheless, scores of places have been rescued from oil and gas drilling, from logging and roading, from strip mining and flooding by well-timed NEPA lawsuits that bought enough time for the legislative act that would finally spare them. A comprehensive list would take pages, but here are a few:

- Little Granite Creek, Wyoming, leased for oil and gas development, leasing blocked by a court, now secure in the Gros Ventre Wilderness;
- Deep Creek, Montana, leased for oil and gas, leases suspended by the court, currently on appeal;
- Mineral King Valley, California, leased for a ski development, now in Sequoia National Park;
- The Cross-Florida Barge Canal, one of the more dramatic examples: ten days after a preliminary injunction was issued President Nixon withdrew the project; the *coup de grace* was administered in the early eighties;
- Admiralty Island, Alaska, contracts signed by Forest Service that would have seen entire island clear-cut, now mostly protected as a national monument;

- Redwood National Park, California, seriously threatened by erosion from logging upslope from park, lawsuit delayed logging until Congress could add critical areas to park;
- Blue Creek, California, Forest Service proposed to build a road and sell timber in remote primitive area important for wildlife and sacred to California Indians. Lawsuit stopped road on environmental grounds, Supreme Court now considering religious-freedom arguments;
- Canaan Valley, West Virginia, proposed for flooding under pumped-storage reservoir, Supreme Court just declined to review decision that spared the valley, which may become a national wildlife refuge;
- Misty Fjords National Monument, Alaska, lawsuit has kept at bay plans for the world's largest open-pit mine (for molybdenum) in the heart of the monument. Eventual fate still unresolved;
- Gore Range Eagles Nest Primitive Area, Colorado, Forest Service proposed logging area without studying its potential as wilderness, area now protected as wilderness.

Another dramatic NEPA victory came in response to suits filed by Gus Speth, a former NRDC attorney and now head of the World Resources Institute, and by Tony Roisman. Speth had challenged the EIS for an experimental fast-breeder nuclear reactor the federal government planned to build at Clinch River, Tennessee. That suit, plus a later one by Roisman that successfully delayed a government plan to commence retrieving plutonium from spent light-water reactor fuel, according to Roisman, "gave Carter the ammunition and the time he needed to kill the plutonium option. Just think how much worse the current wave of terrorism and international tension would be if the world were operating on a plutonium economy."

If the seventies were the golden age of environmental law, what are the eighties? What lies in store? Many environmental lawyers argue that the environmental law business is not so innovative as it was, that pure litigation is getting rarer. It is, they say, being replaced by a complicated hybrid of lobbying, negotiation, compromise, and gentle persuasion.

Fred Anderson, former executive director of the Environmental Law Institute and now dean of the Washington College of Law at American

University, says, "Litigation is less important now. Much has been institutionalized in agency regulations. It gets harder and harder to find big cases to publish and analyze. Now it's a case of trying to push agencies to get them to do what they can, of joining agencies to fend off attacks from industry."

Joe Sax, professor at the University of California's Boalt Hall and author of Michigan's pioneering environmental-quality law, puts it this way: "Environmental law is more mature and less interesting now." And EDF's Tom Graff says, "The Environmental Defense Fund doesn't do as much litigation these days as we used to. We concentrate more on the promotion of ideas and programs dreamed up by economists and scientists. Rather than go to court, we lobby, write reports, court the media."

NRDC's John Adams, however, insists that "litigation is still the single most important part of the environmental movement because the country's so big. The judicial system is centralized. It's the only way to get a grip on the country. Until the laws and regulations are working well, there's still a ton of work to do. Other techniques are catching up with litigation, but there's still a lot for environmental lawyers to do. Can you think of a local fight that has been won without the assistance of a lawyer?

"NRDC seeks leverage situations to get at bigger problems. You can't litigate every oil lease. You keep your eye on the big picture, but you still need lawyers to challenge every new reg."

Sutherland echoes his friend Adams: "Environmental litigation is getting more complex, more sophisticated. There are fewer threshold issues now. But litigation is still essential. It means power for people who don't have economic power. It's one way to fight the political fight. It gives you leverage [that word again], power beyond appearances.

"Add the assets of all the environmental groups and the total would be equal to the assets of a fourth- or fifth-rate oil company you never heard of. Environmentalists can play the political game now: lobbying, publicity, publishing. But they didn't have that power until they started suing."

Assuming that both observations are correct—that the role of litigation is changing and that it's still important—several new themes emerge.

The Private Public-Interest Bar

John Bonine, professor of law at the University of Oregon's Western Environmental Law Clinic, spends much of his time beating the drum for what he calls the "private public-interest bar," which he calls the wave of the future. To some extent it is the wave of the present: A considerable number of lawyers—many with experience in the established groups— are making comfortable livings representing environmental interests.

The informal bar Bonine talks about, however, consists mainly of young, idealistic lawyers outside the big cities who do work for individuals or nonprofit groups for low fees, possible court-ordered attorneys' fees, or for a share of the award, if any, in damage actions.

"A growing number of young lawyers are fashioning careers in environmental law by just hanging out their shingles and doing it. This approach is not for the fainthearted, but it can and does work. Others join a small plaintiffs' firm with like-minded associates. I know of few private lawyers continuously practicing environmental law, however, who work for the large firms of our major cities." He explains that real or imagined conflicts of interest come up in the big firms, putting a severe limit on the contributions of *pro bono publico* help, as vital as that has been and still is.

Bonine's "network" as he characterizes it, consists of these altruistic lawyers who share research and strategies and do not have unlisted telephone numbers. He has compiled a directory that lists 150 such attorneys in the West, and his clinic sponsors an annual conference that brings them together to exchange ideas. Many are heavily into computers. They are needed because more and more situations that demand legal help are localized disputes over waste dumps or pesticide-spraying programs or timber sales. These cases are simply too numerous to be handled by the national environmental law firms, so it is essential to build up a cadre of local or regional lawyers to do the work.

A word of caution, however, is sounded by William Rodgers of the University of Washington Law School. "Hardship," he says, "does not appear to be in vogue. Most law graduates remain committed clerks (even the second professional in the family who has fewer excuses), with the result that environmental cases go begging while lawyers enjoy their security, income, and caution." And David Andrews, a former EPA official now in private practice in San Francisco helping industries comply with federal environmental regulations, warns about inexperienced groups and lawyers clogging up the courts and harassing agencies: "The established groups—Sierra Club Legal Defense Fund, Natural Resources Defense Council, Environmental Defense Fund—are always prepared, always worthy adversaries. Their science is good. They bring meritorious suits. When they get involved you sit up and take note. They don't bring frivolous suits. But as it gets more complex and other groups come in, they could paralyze agencies with fearful, meritless suits."

Bonine replies that the network he is building relies heavily on communication, sharing, and what he calls "mentoring" to ensure high-quality legal work even from those who are new to the field.

Fees and awards

The only way the public-interest bar will thrive will be if lawyers are able to recover fees and costs in successful suits, a principle the courts and Congress tinker with almost constantly. The Supreme Court recently issued several opinions that limit fee awards, which has some lawyers worried.

All might have been different had not the Supreme Court ruled in an Alaska pipeline case that plaintiffs could not claim fees unless the statute—NEPA, in that case—expressly provided for such awards.

"If that decision had gone the other way," Rick Sutherland says wistfully, "it would have brought tens of millions of dollars to people who advocate in behalf of the environment."

As it is, there are a number of statutes, including the Clean Water Act and the Resource Conservation and Recovery Act, that do provide for fee awards. And there is a catch-all statute called the Equal Access to Justice Act that provides for the award of limited fees and costs under other statutes.

Whether such awards can sustain the environmental bar is open to question. Roger Beers's first assignment at NRDC in 1973 was to study potential funding sources for environemntal-law groups. He concluded that attorney fees would never be enough to sustain the big groups like NRDC, EDF, and SCLDF.

They would have to rely on support from foundations and from the public, which they have done with great success. But Tony Roisman thinks it might be better to put more emphasis on suing for damages. "There's no social value in raising money from foundations," he says. "There is in winning judgments against polluters. It scares them and earns you money at the same time."

Alternative Dispute Resolution

Litigation itself has a bad name in some circles, judged to be expensive and time-consuming. Many people argue that "alternative dispute resolution," more commonly known as mediation and arbitration, will displace litigation. Jim Moorman is among its supporters: "The environmentalists have won and don't know it. Everyone's an environmentalist. Most industries have environmental affairs officers. They don't want to be adversaries. They want to find compromises. This will lessen the need for litigation."

Rick Sutherland is not convinced. "Mediation and the like will never be a complete alternative to litigation, because litigation is one of the ways parties get equality of bargaining position. The threat of litigation is one of the factors that has to be included in the issue of whether mediation will ever be successful or not."

The Marketplace

In the early seventies there was a small organization called the Coalition to Tax Pollution. Its mission was to persuade Congress to make polluters pay for the right to pollute—pay so much that they would find it cheaper to clean themselves up. It was, as Jim Moorman put it, a "real nonstarter." Some lawyers think its time may now be coming.

Sutherland suggests a new look at the whole issue of pollution control. "I have a feeling environmentalists may have been riding the wrong horse all along. I'd like to see if the market system would work— taxes and penalties. Do we need a system with EPA the biggest bureaucracy in the world? All these regulations? It deserves a close look.

"You can't junk EPA overnight, of course, but a careful analysis might demonstrate that our massive, cumbersome system of regulations and regulators could be cut back considerably and replaced by a pay-if-you-pollute system."

Here Sutherland is in agreement with his counterparts in industry. Both Ed Frost and Norm Bernstein think the government's environmental bureaucracy is out of control, and Bernstein predicts there may be a backlash in a few years, when the public comes to believe that all these regulations are adversely affecting the competitiveness of American industry and driving jobs overseas. Moorman also agrees: "We erect an elaborate maze and defeat ourselves. The result has raised the costs to industry to no reasonable end."

Proposition 65

But if the public is restless, it seems to be not over too much regulation, but too little enforcement. In what may be the beginning of a new movement, California voters in 1986 took matters into their own hands by adopting a sweeping initiative called Proposition 65 aimed at stopping chemical contamination of groundwater. It enables any citizen to file a lawsuit against anyone discharging chemicals known to cause cancer or birth defects. If successful, the person is entitled to a portion of the penalty assessed against the polluter.

Restoration

Whereas most NEPA suits have been filed to prevent or to improve new projects, Fred Anderson suggests that some of the focus is shifting now to litigation that will force restoration of damaged places, mitigation of proposed projects, and rescue of endangered animals. "The only lawsuit I prosecuted myself concerned Otter Creek, which had been destroyed seventy years earlier. We thought it had again become a wilderness, and first the court, and then the Congress agreed. Give me a ruined marsh and seven years and, absent toxins, I'll give you back a stable ecosystem, even a rookery."

Cleanup and restoration are the approach of Superfund, which got a great shot in the arm in 1987 when Republican legislators persuaded a reluctant President Reagan to sign it, despite its $9 billion price tag. It has a sweeping citizen suit provision, and is sure to spark considerable litigation, if, as expected, EPA fails to stay on schedule with its cleanup efforts.

Efforts to advance recycling of materials or to turn garbage to energy may well lead to litigation as well, as concern over the by-products of trash incinerators grows.

A Return to Common Law

It seems increasingly likely, as Bill Butler, former general counsel of the National Audubon Society, puts it, that we will see "the private plaintiff's bar become more interested in bringing private 'toxic torts' suits. Several large recoveries have been won in cases like the one in Woburn, Massachusetts [where a discharger of toxic wastes into groundwater recently reached a multimillion-dollar settlement with nearby residents who had sued, claiming their wells had been dangerously polluted]. Terrible accidents like the one in Bhopal and litigation like the Agent Orange case on behalf of servicemen have also tended to educate the private bar about the possibilities of private toxic-tort suits. Environmental groups are now more involved in enforcement actions where fee recoveries and penalties are available." And one of these days, if someone can finally win a liability suit against a tobacco company, the floodgates will open.

Mike McCloskey, chairman of the Sierra Club, adds that "as a result of asbestos and Bhopal and Agent Orange, the chemical industry and others are concerned about their exposure to huge damage suits. Think if the worst had happened at Three Mile Island or at Institute, West Virginia.

The concern is both with catastrophes and with the effects of long-term exposure and delayed cancers. This could easily lead to bankruptcies. So the key is insurance, and the insurance industry says it can't—or it won't—insure against this sort of problem. The nuclear experiment was protected in advance by the Price-Anderson Act, and there might be an attempt to enact a Price-Anderson sort of law for the chemical industry."

Roger Beers has a similar warning: "There may be a move to 'reform' the tort system, to the benefit of the corporate sector. It would complete the Reagan revolution—eviscerate the regulatory agencies saying that it's up to states and local jurisdictions to provide that kind of relief. Then turn around and take away the only kind of monetary relief left to citizens."

The Bench

In a speech to the Knights of Columbus on August 6, 1986, Ronald Reagan said that by the end of his second term he will have appointed 45 percent of all federal judges. The point of the remarks was that both Reagan and the Knights (described by *The New York Times* as "the world's largest Catholic laymen's group") oppose abortion and that the president has been packing the courts with antiabortion judges. He makes no apology, in other words, for applying an ideological test in choosing his judicial appointees.

This, of course, was well before the unsuccessful appointments of Robert Bork and Douglas Ginsburg to the Supreme Court, which may embolden senators to take a hard look at judicial nominations made during the last year of the Reagan administration.

Many environmental lawyers, nonetheless, are alarmed at what the Reagan appointments may hold for the future of environmental litigation. Others take the view that it is impossible to predict how judges will act and that some will likely be welcome surprises.

"Environmental cases are the most political of all," Beers insists. "People consider the Chicago Seven trial and others of that ilk to be political. But truly, the most political cases are environmental cases. Often, as important as the law being sued under, or the specific situation at hand, is the political attitude of the judge. If the judge thinks what you want is trivial or will impede economic growth, you may not get very far if the law is ambiguous."

Don Harris, president of the Sierra Club Legal Defense Fund, believes, "The District Courts are where it's worrisome. The laws are fairly clear, and broad principles are fairly well established. But when judges look at facts they can really screw you. If a bad decision is made on the facts, an appeals court is unlikely to review the decision."

A brief digression may be appropriate here. Lawyers, particularly environmental lawyers, do not much like facts. A case that relies heavily on facts—how much of a given contaminant did the industry discharge on a given day, under what weather conditions, with machinery that did or did not malfunction, for example—is complicated and very expensive to prosecute, owing to the need for research, expert witnesses, technical consultants, and perhaps computer time. Most environmental lawyers much prefer cases that turn on disputes over law: Which law applies to a given case? Or, given the following facts, did the defendant actually break the law? Did the agency involved abuse its discretionary powers? What did Congress really mean when it passed the law?

The optimists say that there is nothing incompatible between being a conservative and being an environmentalist, therefore conservative judges are not to be dreaded. As Fred Anderson observes, "Nixon appointed some environmentalists. Reagan may have, too, without knowing it." Bill Futrell, president of the Environmental Law Institute, says, "We won't know for ten years how the appointments will turn out. A number of them could be excellent. The standards of competence are generally quite high. And we don't know what issues will be before the courts ten years from now."

Durwood Zaelke of the Legal Defense Fund agrees. "As long as there are competent lawyers appointed as judges we're okay. Environment cuts across left-right ideology. No matter who the judge is, there's always a way to win. I'd hate to think we have to depend on judges to win."

However the courts turn out, there is little doubt the law will continue to be one of the environmental movement's most potent tools. It works, and it rewards its practitioners. Most of the young lawyers that started practicing environmental law nearly twenty years ago are still at it, which means both that there is a seasoned cadre of environmental attorneys in the field—and that it's a damnably tough field to break into. Karin Sheldon of the Sierra Club Legal Defense Fund explains why: "Environmental law is exciting to practice, the best work there is. Many of us are relics of the sixties generation that believed that 'the system' could be reformed through individual and collective effort. While our goals remain valid, we have become more realistic about the extent of our power and influence."

Two more points bear mentioning. One, environmental litigation is only as good as the political sophistication and organization that back it up.

For example, in 1973, the West Virginia Division of the Izaak Walton League and other conservationists decided to challenge the Forest Service's widespread practice of clear-cutting in a case involving a proposed clear-cut on the Monongahela National Forest. NRDC attorneys Angus Macbeth and Toby Sherwood developed the novel argument that clear-cutting was a violation of the 1897 statute that established the Forest Service. Under the law, the foresters had been authorized to sell "dead, mature, and large growth of trees . . . for the purpose of preserving the living and growing timber and promoting the younger growth on national forests." The Forest Service gamely argued that "large growth of trees" meant a sizable stand or grouping of trees and that "mature" meant economic, rather than biological, maturity. The court rejected the Forest Service's argument and held that clear-cutting was against the law. The government appealed and lost again.

But the tale did not end there. The Forest Service and timber industry turned to Congress for relief, arguing that times had changed and that the old law had to change as well. The outcome was the National Forest Management Act of 1976, which despite the best efforts of conservationists, makes clear-cutting legal in many circumstances.

Much the same thing happened in another 1973 case in which the Wilderness Society and others sought to block construction of the Trans-Alaska Pipeline. NRDC attorney Tom Stoel, Jr., argued that the proposed pipeline violated the Mineral Leasing Act. The Supreme Court agreed, but Congress quickly exempted the pipeline from the law. Thus, a consideration frequently weighed when a suit is contemplated is whether a victory will last longer than the time it takes industry lobbyists to complain to Congress.

Second, the world gets ever more complicated. As David Sive notes, "The environmental movement has passed from the Messianic phase to a more mature phase. It's harder to tell the good guys from the bad guys. When Martin Luther King, Jr., sat down at that diner in Tuscaloosa, the issue was as clear as the sky over Arizona. Now civil rights issues include affirmative action, quotas, and so on. They're much more complicated. The same is true of environmental questions."

Complicated or not, these issues and the lawyers who thrash them out in the courts are here to stay. As former Secretary of the Interior Stewart Udall observed, "Today, environmental law is part of the warp and woof of American life. It impacts policy making at all levels of government, and the environmental law organizations exert a powerful influence on lawmaking in Washington—and guide the enforcement and implementation of the nation's environmental laws."

David Brower, who started in the conservation movement more than fifty years ago, says simply, "Thank God for the lawyers. There have been a lot of places saved through the courts."

4

The Ecophilosophers

Peter Borrelli

IN SURVEYING the state of environmentalism, I am struck by its diversity. The organized movement comprising national and international organizations—themselves as diverse as the Natural Resources Defense Council and Greenpeace—is but the most visible part of a global awakening and transformation.

In 1948, the noted astronomer Fred Hoyle anticipated the current situation when he wrote, "Once a photograph of the Earth, taken from the outside is available . . . a new idea as powerful as any other in history will be let loose." That photograph is now available, and with it the understanding that the world we live in is both finite and interconnected.

More recently, Alvin Toffler described the eighties as the dawn of a new era of crisis and opportunity regarding the ultimate survival of the planet. "Humanity," he wrote in his best-seller *The Third Wave* (1980), "faces a quantum leap forward. It faces the deepest social upheaval and creative restructuring of all time. Without clearly recognizing it, we are engaged in building a remarkable new civilization from the ground up."

Still others, notably the new-new environmentalism founded on the principles of global interdependence, describe this transformation as a paradigm shift: from the industrial-age world view that has dominated our social and economic institutions, to an ecological world view.

If you were to poll the leaders of the national environmental organizations, most would express sympathy for, and in some instances profound understanding of the commandments of ecological wisdom. Environmentalism has given new meaning to the biblical injunction to be our brother's keeper; indeed, the environmental cause is the cause of life. However, most would also subscribe to Edmund Burke's belief that "nobody made a greater mistake than he who did nothing because he could do only a little." For the environmentalists concerned *today* about the destruction of tropical forests in Brazil or the acidification of Adirondack lakes, waiting for the paradigm shift may be like waiting for Godot.

The old-old and new-old movements are grounded in a kind of radical pragmatism. That view is challenged increasingly by a growing number of splinter groups that prefer the radical to the pragmatic, whether or not they can effect immediate change. The monkeywrenchers reported on in part two of this series are among them. Others include deep ecologists, bioregionalists, and the Greens. Each of the groups defies cubbyholing, as there is much overlap between and among such groups and even with the traditional organizations, but a bit of history helps to place them.

•

We are coming to recognize as never before the right of the nation to guard its own future in the essential matter of natural resources. In the past we have admitted to the right of the individual to injure the future of the republic for its own present profit. The time has come for a change. As a people we have the right and the duty, second to none other but the right and duty of obeying the moral law, of requiring and doing justice, to protect ourselves and our children against wasteful development of our natural resources, whether that waste is caused by the actual destruction of such resources or by making them impossible of development hereafter.

With this epochal declaration, made spring 1908 at a White House conference on conservation, Theodore Roosevelt gave formal birth to the *conservation* movement in America. Gifford Pinchot, his chief advisor on natural resources, later described the occasion as the birth of "a world movement."

Roosevelt's brand of conservation set the course that others would follow for decades. Its focus was responsibility and restraint in *managing* natural resources, and its opponent *within* the camp was preservationism (led by John Muir), which favored *protecting* the earth from the hand of man. The tension between management and preservation is present to this day in both natural resource agencies and the environmental movement itself.

The conservation/preservation split reflects far more than disagreement over humanity's responsibility toward the land: It is rooted in fundamentally different perceptions of humankind and nature. Roosevelt and Pinchot shared an anthropocentric world view, in which man is the measure of creation. Muir's world view was biocentric; man is but one form of life, one part of creation. Muir once asserted, writes biographer Frederick Turner, that "he had 'precious little sympathy for the selfish propriety of civilized man,' and if there should ever chance to occur a war between man and the wild beasts, 'I would be tempted to sympathize with the bears.' " It is an idea shared by a great many people today.

A quarter-century later, the general public, mired in the depths of economic depression, still was blind to the wisdom of conservation (not to mention preservation), but the theme of restraint continued to resound in the policies and programs of the New Deal. As Franklin D. Roosevelt declared in his Commonwealth Club speech given in San Francisco in 1932, the federal government had to restrain big business in the name of the public interest. A principal architect of FDR's economic policies and member of his famous "brain trust" was Adolf A. Berle, coauthor of the seminal book, *The Modern Corporation and Private Property.* An advocate of "enlightened administration," more popularly known as good government, Berle insisted that "the modern corporation serve not alone the owners or the [managers] but all the society." (Berle's son, Peter, is currently president of the National Audubon Society.)

By the 1960s, the public had awakened to the rapidly deteriorating state of the nation's (and to a much lesser extent, the world's) environment. And with the celebration of Earth Day in the spring of 1970, the modern *environmental* movement was born, charged with confidence and poised to do battle with *the system*. There was not a great deal of talk about changing the system. The notion of user restraint again guided a string of federal command-and-control laws.

The environmental movement of the late sixties and early seventies encompassed a wide range of political and social interests, bringing together the old movements and young activists inspired by the civil rights, women's, and antiwar movements. The widespread hope of progressive change led to the creation of such diverse groups as NRDC, Nader's Raiders, and Greenpeace. Through a concerted effort to set aside differences, the movement more or less functioned as one, firing the public imagination and growing exponentially.

But the sense of common purpose has unquestionably lost its hold on both individuals and organizations. One explanation is that "the great undertaking" has not measured up to expectations; environmental problems, especially those measured on a global scale, overshadow accomplishments. The environmental problem has become part of a much larger one touching on everything from modern technology to social justice to ethics and religion. Equally important, the good-government-vs.-big-business model, which has influenced conservation and environmental policy since the Progressive era of Teddy Roosevelt, has lost its unifying appeal. The "enemy of the people" may be not the individual manufacturer of a polluting automobile, but the economy that requires its production, the consumer who craves it, and the government that regulates it. So say the new-new environmentalists.

Michael McCloskey, chairman of the Sierra Club and from 1969 to 1985 its executive director, notes, "While the potential for such splits was evident at the outset, the initial sense of common investment in a great undertaking may have acted as a restraining force. Now, however, some of that self-discipline may be breaking down as more militant forces emerge."

•

Deep ecology is the most influential new way of interpreting the environmental crisis. Its principles, enunciated in 1972 by Norwegian philosopher Arne Naess, are defined principally in opposition to the established movement, which it terms "shallow" ecology: anthropocentric and utilitarian. In this view, shallows encourage the destructiveness and wastefulness of industrial society even as they seek to reform it. For example, deep ecologists argue that environmental regulations based on emissions standards and tolerances represent licenses to pollute. Thus, NRDC, National Audubon, the National Wildlife Federation, and other such groups are considered shallow by most deeps.

Naess's ideas have been developed in the United States by sociologist Bill Devall and philosopher George Sessions. (Both are Californians and active in the Sierra Club, despite its "shallowness.")

Deep ecology extends the ecological principle of interrelatedness to virtually every aspect of our daily lives. Human and nonhuman species are viewed as having inherent and equal value, from which it follows that humans have no right to reduce the natural diversity of the earth, either directly or indirectly. Direct actions include such things as agriculture, mining, forestry, and technology. Indirect actions include economic and social policies that impinge on other human or nonhuman life forms, as well as population growth, which is viewed as an impediment to "the flourishing of human life and cultures."

Deep ecologists do not specify what the optimal human population of the world should be, but Naess has suggested an unimaginable figure of 1 billion, roughly equal to the total world population in 1800. In 1986, Naess wrote with all seriousness in the journal *Philosophical Inquiry:* "It is recognized that there must be a long range, humane reduction through mild but tenacious political and economic measures. This will make possible, as a result of increased habitat, population growth for thousands of species which are now constrained by human pressures."

John Muir, who may have been this country's first deep ecologist, summed up the biocentric (some might say misanthropic) perspective in these words:

> Pollution, defilement, squalor are words that never would have been created had man lived conformably to Nature. Birds, insects, bears die as cleanly and are disposed of as beautifully. . . . The woods are

full of dead and dying trees, yet needed for their beauty to complete
the beauty of the living. . . . How beautiful is all Death!

According to Sessions and Devall, the goal of deep ecology is to
achieve universal ecological consciousness "in sharp contrast with the
dominant world view of technocratic-industrial societies which regards
humans as isolated and fundamentally separate from the rest of nature,
as superior to, and in charge of, the rest of creation." McCloskey's response:
"While many of the aims of deep ecologists are appealing as ideals, it is
not clear how far or fast they want to go in pursuit of their ideals, nor
whether they really want to engage in the process of real-world change."

Indeed, deep ecology is less a movement than a bundle of ideas
held to a greater or less degree by a heterogenous grouping of organizations and individuals. The most visible and best organized are the monkeywrenchers, Earth First!, who take direct action in a variety of wilderness
and forestry controversies in the West by arguably nonviolent means.
(Naess, incidentally, is the author of two studies on Gandhi and reportedly
engaged in nonviolent resistance against the Nazis.) Deeps also are found
on campuses and in local grass-roots organizations, where their effectiveness is hard to measure.

McCloskey acknowledges that an attitudinal survey of five major
environmental groups (Environmental Action, Environmental Defense
Fund, National Wildlife Federation, Sierra Club, and the Wilderness Society), conducted by Resources for the Future in 1978, suggests that about
19 percent of their members hold views "that might be associated with
the deep ecology movement."

•

Bioregionalism resembles deep ecology in adopting a radically different world view but is more catholic and tied to a different social base.
It is an outgrowth and resurgence of the sixties back-to-the-land movement,
and in the spirit of E. F. Schumacher's *Small Is Beautiful,* looks for inspiration to cultures and life styles of indigenous peoples and dwellers
in the land such as the Amish. Currently, there are more than 100 bioregional groups from Oregon to Maine, which informally exchange information and ideas. Since 1984, there have been two North American

Bioregional Congresses attended by representatives from thirty or more states, several Canadian provinces, and Native American tribes.

The term bioregionalism was coined in 1976 by Peter Berg, who, with his wife Judy Goldhaft, runs a bioregional clearinghouse in San Francisco's Mission District called the Planet Drum Foundation. (The drum figures in shamanic ritual.) During the sixties, Berg and Goldhaft were among the founders of the Diggers, an organization that offered refuge to thousands of young people descending on Haight-Ashbury. It was also during this period that they made a cross-country tour of back-to-the-land communes.

The underlying premise of bioregionalism is that the environmental crisis begins at home and revolves around individuals' perceptions of their place in the world. Being a bio requires bearing witness. "All the planet-wide pollution problems originate in some bioregion," says Berg. "It's the responsibility of the people who live there to eliminate the source of them. If you strike a five-finger chord—by promoting more community gardens, more renewable energy, accessible public transportation, sustainable agriculture—and you are serious about it, rain forests will stop getting chopped down in direct proportion."

Poet Gary Snyder, whose Pulitzer Prize-winning *Turtle Island* captures the spirit of bioregionalism, has said that he saw himself "living on Planet Earth, on Turtle Island, in Shasta Nation. I think in terms of where the plant communities shift, and know where the rivers reach better than I know the highways now. It's a wonderful way to see the world, and it'll outlast anyone's local political boundary."

Snyder has a bioregional quiz for guests. "Where does the water you drink come from? What is the soil series where you live? Name five edible plants in your region and their seasons. Where does your garbage go?"

Unlike the back-to-the-landers who tended to isolate themselves in self-supporting communes, bioregionalists for the most part live within existing communities where they carve out simplified life styles and become involved in regional land-use issues. While some try to live off the land as farmers, most are wage earners in a variety of trades and professions. In many respects, they are the opposites of yuppies. Young, well-educated, and potentially upwardly mobile, they have chosen to make their living where they want to live rather than the other way around.

In the Sierra Nevada foothills where Snyder lives, he and others have joined long-time residents in electing new planning commissioners concerned about controlling growth. They also are involved in the U.S. Forest Service's management plan for the Tahoe National Forest.

In Northern California's Mattole River area, friends of the Bergs are involved with a watershed bioregional council that has been working with local residents to restore the native king salmon population, destroyed by logging operations that had silted the river.

When drinking-water contamination got out of hand along the Missouri-Arkansas border, a group of people got together to form the Ozark Area Community Congress (OACC), which promotes the use of compost toilets. In 1980, OACC called the first large gathering of the "ecological nation," modeled after tribal consensus as practiced by the Iroquois Federation.

OACC's activities soon attracted attention in Kansas, where it inspired formation of the KAW Council based in Lawrence. Among its founders is thirty-year-old Kelly Kindscher. In 1983, he took a 690-mile "walk" from Kansas City to the foothills of the Colorado Rockies "in the tradition of the Indian vision quest" to get to know his prairie bioregion. KAW now has about 350 members, publishes a newsletter, and is involved in agricultural issues.

A strictly grass-roots movement, the idealism of bioregionalists is balanced against their pragmatism and willingness to work for change within the communities where they live. Many belong to established environmental organizations and are trying to get the Church more involved in environmental issues.

Despite their willingness to work incrementally toward change, most bioregionalists believe the trend toward ecological destruction will not be reversed until there is a spiritual awakening. Father Thomas Berry, director of the Riverdale Center for Religious Research in New York and active with the Hudson Bioregional Council, states, "If we do not alter our attitude and our activities, our children and grandchildren will live not only amid the ruins of the industrial world, but also amid the ruins of the natural world." What is needed, he says, is a "treaty" or spiritual bond between ourselves and the natural world similar to God's covenant with creation after the Flood (*I set my rainbow in the cloud, and it shall be a sign of the covenant between me and the earth.* — Genesis 9:13). Such a treaty would be based on the principle of mutual enhancement. "The [Hudson] river and its valley," he writes, "are neither our enemy to be conquered, nor our servant to be controlled, nor our mistress to be seduced. The river is a pervasive presence beyond all these. It is the ultimate psychic as well as the physical context out of which we emerge into being and by which we are nourished, guided, healed, and fulfilled."

Berry's writings have attracted considerable attention among clerics, scientists, and environmentalists who feel that the Church has turned its back on the biological sciences and belittles "nature worship." Brian Swimme, a California physicist who works with the Institute for Cultural and Creation Spirituality, credits Berry with having developed a "functional cosmology" that goes beyond both science and theology. By combining a contemporary biological view of life with a sense of mystery about the universe, Berry has given new meaning to spiritual devotion and religious responsibility.

•

Though not actually a movement, the Gaia hypothesis complements the thinking of many of today's environmentalists. First proposed in 1972 to British chemist James Lovelock, who had worked with the NASA team investigating the possibility of life on Mars, it holds that "the evolution of the species of living organisms is so closely coupled with the evolution of their physical and chemical environment that together they constitute a single and indivisible evolutionary process." The idea was drawn from Lovelock's observation that despite being composed of an unstable mixture of reactive gases (such as oxygen and methane), the atmosphere for the entire 3,500 million years since life began has remained remarkably stable. From this he hypothesized that "living matter, the air, the oceans, the land surface, were parts of a giant system which seemed to exhibit the behavior of a living creature." Hence, Gaia, Greek goddess of the earth.

The idea is not new. It surfaced during the nineteenth century when the German geographer Carl Ritter postulated a galvanic force in nature by means of which its separate parts communicated. More recently, the hypothesis has been extended by the controversial work of Boston University biologist Lynn Margulis, who argues that symbiosis and cooperation among all organisms have been integral to successful existence.

Whether metaphor or reality, Gaia is in serious trouble. The dynamic balance of the biosphere is being radically disrupted by human activities such as the destruction of tropical forests, fossil-fuel consumption, and the release of chlorofluorocarbons.

•

While there is a widespread belief these days that caring for the earth requires fundamental and even radical change, it is not at all clear that this must extend to electoral politics. In the United States, the environment has been a bipartisan issue with vast public support. As former Senator Gaylord Nelson observed in part one of this series, when he went to Washington in 1963, "There were not more than five broad-gauged environmentalists in the Senate." When he introduced a ban on DDT during his freshman year, he could find only a single sponsor for a companion bill in the House. Were such a bill introduced today, however, the list of sponsors would fill the title page. The environment is now called America's issue, which at the very least means that there is some kind of consensus about the importance of environmental issues, if not their solution.

In the democratic coalition governments of Europe, where representation in parliament is based on the percentage of the national vote won by each party, it has been another story. There, young activists and elements of the middle class have coalesced around the ideas of deep ecology and the goal of achieving *Ökopax* (eco-peace), campaigning for a nuclear-free Europe and against environmental abuses. In 1979, the West Germans formed a coalition: *die Grünen,* or the Greens. By 1983, the Green Party had won 5 percent of the national vote (the minimum share of the vote required to win seats in the Bundestag), and twenty-seven seats in the Bundestag.

The Greens' astonishing entrance onto the political stage was immediately followed by infighting among ecologists, counter-culturalists, moralists, feminists, antinuclear activists, and Marxists—each insisting that his or her prescription for the world's ills was most effective. While demonstrating the party's broad base of support, the infighting, nevertheless, has diverted its energy. Less than a year after their entry into parliamentary politics, one of the most articulate of the Greens, a former general in the West German army named Gert Bastion, quit his seat in the Bundestag in disgust. A few months later, Petra Kelly, a founding member, was thrown out of the Bundestag by her own party for ignoring her Bavarian constituents.

Still, by not trivializing their agenda and constantly seeking political compromise, the Greens have survived. Buoyed by protests against a proposed nuclear reprocessing station in Wackersdorf, public reaction to Chernobyl, and progress toward nuclear arms reductions, the Greens won 8.3 percent of the popular vote and forty-two seats in last year's national election.

The Greens are also active in Italy, where they hold fifteen seats in Parliament; Belgium, where they have formed a coalition with the Social Democrats; and in the United Kingdom, where they fielded 133 candidates in last year's general election and won 1.36 percent of the vote. At last count, there were Greens in seventeen countries, including the United States.

Soon after the West German Greens' first electoral victory, Charlene Spretnak, a Berkeley activist and lecturer on spirituality and feminism, and Fritjof Capra, author of *The Tao of Physics* and founder of an ecological think tank called the Elmwood Institute, teamed up to write *Green Politics*. It called for a U.S. Green movement based not on the German model but on the spiritual insights of deep ecology. The idea was that ecological problems of America were not the direct fault of capitalism but of consumerism caused by an emptiness of spirit.

Though deep ecology disavows central control, Spretnak initiated a loose-knit grass-roots network called the Committees of Correspondence (as during the American Revolution), organized at a meeting of activists in St. Paul, Minnesota, in August 1984. They adopted a platform consisting of "ten key values": ecological wisdom, grass-roots democracy, personal and social responsibility, nonviolence, decentralization, community-based economics, post-patriarchal values (feminism), respect for diversity, global responsibility, and concern for the future. Since then, about seventy-five Green groups have been formed.

In the summer of 1987, about 1,500 Greens gathered in Amherst, Massachusetts, for their first national conference. The stage was set for the Greening of America, but by lunchtime of the first day, the meeting had dissolved into an unhappening, as shouting matches broke out among deep ecologists, feminists, animal liberationists, anarchists, antimilitarists, monkeywrenchers, and graying SDSers. There was little talk of national organization or coalition building.

The Greens face numerous problems in the United States, not the least of which is the entrenchment of a two-party political system. While there is room for new, even radical environmental thinking that challenges the underlying assumptions of modern technology and industrial societies, the Greens must cope with the reality of environmentalism already having been absorbed by American politics. The likelihood of the Greens becoming a third national party is extremely remote, but the possibility of their stimulating effective local coalitions of minorities, women, workers, neighborhood groups, and environmentalists involved in every-day challenges such as toxic exposure and community planning and development is very real.

In New England, the Greens' first political success came in 1985 in New Haven, Connecticut, where they received 10 percent of the vote, edging out the Republicans as the city's number two party. A local school teacher running for mayor on the Green ticket was defeated, but not without adding interest to the election. Their platform included support for inner-city victory gardens, a program to remove asbestos in homes, public ownership of the local electric utility, and conversion to renewable energy sources. Five black Democrats have since announced their intent to run for aldermen as Greens.

Elsewhere, Greens were instrumental in pressuring the Philadelphia City Council to pass the country's first law mandating trash recycling, and in defeating a luxury condo project in Burlington, Vermont. In Kansas City, the Greens are involved in direct food marketing, buying from area farmers and donating the produce to distribution centers for the homeless. The Lone Star and Longhorn Greens of Austin, Texas, halted construction of a power line through an environmentally sensitive area and successfully ran a candidate for city council. The San Francisco Greens have begun an educational campaign on toxic substances in the home, while their neighbors in the East Bay Green Alliance are restoring a salt marsh.

An important faction among the Greens is social ecology, which holds that environmental problems are rooted in social conditions. The leading spokesman, Murray Bookchin, is a veteran of the old and new lefts and director emeritus of the Institute for Social Ecology in Plainfield, Vermont. Bookchin takes sharp issue with those who "deify" nature.

> Deep ecology has parachuted into our midst quite recently from the Sunbelt's bizarre mix of Hollywood and Disneyland, spiced with homilies from Taoism, Buddhism, spiritualism, reborn Christianity, and, in some cases, ecofascism, while social ecology draws its inspiration from such outstanding radical decentralist thinkers like Peter Kropotkin, William Morris, and Paul Goodman . . . who have advanced a serious challenge to the present society with its vast hierarchical, sexist, class-ruled, statist appartus and militaristic society.

"We must achieve not just the reenchantment of nature but the reenchantment of humanity," Bookchin told the Amherst conference. He argued that the deep ecologists view humanity as "an ugly 'anthropocentric' thing—presumably, a malignant product of natural evolution—that is 'overpopulating' the planet, 'devouring' its resources, destroying its wildlife and the biosphere.

"Our wholesale condemnation of technology," he continued, "is a condemnation of some of the best achievements of mankind. The problem begins with our hierarchical society in which growth and wealth measure progress and culture does not."

Bookchin has likened monkeywrenchers like Earth First! to Nazi thugs, calling its founder, Dave Foreman, a "macho mountain man." He believes that the deep ecologists' preoccupation with world population smacks of elitism and racism. He argues, for example, that Thomas Malthus (1766–1834), the English economist and clergyman who warned that "human population growth would exponentially outstrip food production," was not a prophet but "an apologist for the misery that the Industrial Revolution was inflicting on the English peasantry and working classes."

Dan Chodoroff, also of the Institute for Social Ecology, echoed Bookchin's remarks. "We need not a mindless unity with the rest of nature, but an increased mindfulness . . . a humanistic tradition which looks at the vast fecundity and tremendous diversity of relationships in nature, which need to be developed and applied by human beings."

•

We began by observing that environmentalism is at a crossroads. This is not necessarily a gloomy state of affairs, but one that requires reflection and regrouping. The biosphere is under greater stress than at any other time in human history, and both our physical and psychic survival require that we act on that knowledge.

Ecology is not only a science, but an ethical confirmation of the wisdom of the great religions. Wes Jackson, a brilliant thinker in the field of sustainable agriculture, reminds us in *Altars of Unhewn Stone* of God's words to Moses after he had delivered the Ten Commandments: *If you make me an altar of stone, you shall not build it of hewn stone; for if you use your tool on it, you have profaned it.* —Exodus 20:25. "The scripture must mean," writes Jackson, "that we are to be more mindful of the creation, more mindful of the original materials of the universe than of the artist."

As we approach the end of this century—and the beginning of the next—it is appropriate to consider if we have made the best use of our atomic- and space-age knowledge. Environmentalists, in particular, should be asking if they have sufficiently advanced their cause. There are many who believe we have not, either because our vision is too dim or our actions too timid.

The danger is not that the cause will suffer for having more than one philosophical or political idea, but that it will fail to have any. It has been noted by many pollsters, for example, that public support for environmental issues has grown over the past twenty years and is now in excess of 80 percent. However, when the same pollsters attempt to measure the salience of prominence of environmental issues in the public mind, they come up with startlingly different results. Virtually nobody (less than 1 percent) considers the environment the number one problem facing the nation, and seldom is it a key factor in national political campaigns. Although this discrepancy says something about the limitations of opinion polls and shallowness of the political process, it also gives some cause for alarm. Either the seriousness of the present situation has not sunk in, or else it has, and it's every man, woman, and child for himself in a mad dash toward yuppie heaven.

The English writer J. A. Walter, in *The Human Home,* asks:

> How are Americans to restore the environment to its proper position as an essentially political issue, over which reasonable people will disagree? . . . How can we regain the attitude of those earlier ages which saw the natural world as pointing to the divine without itself being divine? How can we cherish our environment without making a fetish of it?

The answer to such questions lies in recognizing that pollution, plunder, consumption, and waste for the most part are the consequences of political and economic conditions that also account for much of the human misery and strife in the world. This will require a broader, more comprehensive vision on the part of national environmental groups and a pragmatic, more constructive strategy on the part of groups seeking radical change. More important, it will require greater commitment and courage on the part of all environmentalists to challenge the present direction of economic policy and politics. Our choice is between living and surviving.

Part II

Perspectives

5

A View Toward the Nineties

William K. Reilly, Jr.

AS THE UNITED STATES approaches the 1990s, two factors above all characterize the state of affairs in environmental policy:

- First, the country faces an array of environmental problems even more daunting than pollution crises of the past generation.
- Second, current policies and institutions, having addressed the easiest matters, seem increasingly unable to deal with these emerging problems.

In short, the programs our country has created to tackle environmental problems are not up to the job ahead. We must critically examine and then change our policies and institutions, or prepare to face serious threats to public health, the environment, and the national economy.

To be sure, the United States does not now face an environmental crisis. Progress continues in abating some kinds of pollution problems in

87

some places, and in the short haul no impending disasters can be predicted from a failure to address any of the lengthy list of environmental issues. Looming ahead, however, is a set of complex, diffuse, long-term environmental problems portending immense consequences for the economic well-being and security of nations throughout the world, including our own. These problems challenge our country's leadership to establish a new course for U.S. environmental policy at home and abroad.

Before looking toward the future, we must acknowledge that there has been continuing progress over nearly two decades in improving some aspects of environmental quality. Levels of particulates, sulfur dioxide, nitrogen dioxide, and other air pollutants, for example, are trending downward in comparison with levels that were prevalent a decade or more ago. While ozone levels also seem to be dropping, this pollutant remains especially difficult to control; it is anticipated that many communities will not achieve compliance with federal health-based standards for exposure to ozone by the end of 1987, as the law requires. Millions of Americans are affected.

In water quality, too, there are some signs of improvement. Monitoring indicates that levels of fecal coliform and dissolved oxygen are decreasing in some bodies of water. People are swimming and fishing in rivers that once presented hazards to health. Some signs of improved quality are appearing in the Great Lakes—especially Lake Erie, once virtually written off as a dying body of water.

Toxic wastes released into the air or water or disposed of on land still present health and environmental risks, but at least some consideration is now being given to the effects of *new* chemicals *before* they enter commerce. In the natural resource area, substantial amounts of land have been protected in parks, wilderness areas, and wildlife refuges. For some endangered wildlife—the whooping crane and the peregrine falcon, for example—the threat of extinction has diminished.

In other words, determined actions to improve environmental quality during the eighteen years since Earth Day have yielded positive results.

As significant as these developments are, perhaps the most important step forward has been the broad recognition by the U.S. public that the relationship between people and their natural surroundings, between human well-being and a healthy environment, must be a matter of ongoing concern. More than any other factor, the evolution of public attitudes, manifested in deep-seated political support for environmental programs, underlies the environmental progress this country has made.

Also encouraging is the fact that many of the worst effects of the minimalist federal environmental policies prevalent in the early 1980s have been reversed. Integrity and good management have been restored at the U.S. Environmental Protection Agency (EPA). The rhetoric of confrontation, so prevalent during the early Reagan years, has softened somewhat. The State Department's Agency for International Development has taken an active role in U.S. efforts to protect the environment of developing countries. Under U.S. government pressure and new leadership, the World Bank, too, has taken major steps to improve its environmental performance in the developing world.

To be sure, bastions in federal and state governments and elsewhere continue to preach the discredited line that the market, in and of itself, will safeguard the environment. More sensible is the widespread and growing recognition of the need for monitoring, analysis, and intervention, as well as for effective partnerships among different levels of government, to improve and protect the environment and to assure continuing supplies of the resources we need to survive and prosper.

Notwithstanding evident progress, the need for environmental action is at least as great as it has ever been. Problems long recognized remain unsolved, and new ones continually appear. Public understanding of environmental threats lags well behind reality, and political consensus on how to meet new needs is not apparent. Past successes belie a growing incongruity between where the problems are greatest, and where priorities and money for cleanup are directed.

If citizens or policy makers or corporate leaders in the early 1970s once believed that the primary job in environmental protection would end when the air and water were cleaned of a handful of known contaminants, this belief has been dispelled. With greater knowledge and with heightened public awareness, the catalog of environmental problems confronting the country has grown, making obsolete the notion that a simple checklist of environmental problems can be dealt with once and for all.

The following examples suggest the range of problems and the difficulties of fashioning and implementing responses.

● *Although current air quality programs have helped clean the air in the vicinity of emission sources, many pollutants escape and are carried much longer distances in the upper atmosphere than previously thought, before they fall to earth.* Acid rain is the most publicized example. Another instance is the large amount of toxic chemicals, like PCBs in the Great Lakes, that come from pollution settling out of the air. The disparity between the geographic scope of environmental problems and the jurisdictions of the governments that must deal with the problems is becoming an acute weakness in efforts to improve environmental quality. The jurisdictions creating the problems have no incentive to do anything about them, while those jurisdictions on the receiving end have no power over far-off sources.

● *Indoor air pollution is another serious problem current programs do not address.* Most people spend most of their time indoors, and pollutants inside are found in much higher concentrations than levels outside. One source, smoking, is notoriously difficult to curb. Other sources of indoor pollutants, such as materials or building systems widely used in new construction throughout the country, may not have ready substitutes. The national debate has not really begun about how to respond to the discovery that indoor air pollution often far exceeds minimum health standards for air outdoors.

● *Growing conflicts surround the allocation of fresh water supplies.* Especially in the West, state water allocation systems established in the last century have allocated all the available water, and then some, to uses such as irrigation, ranching, and mining. Today, as values are changing, Indian tribes, urban populations, and others are asserting new claims on fresh water for recreation, industrial expansion, and preservation of aquatic environments. Mechanisms that can adjust water allocations to the emerging situation are woefully inadequate. The shift from an era of water development to an era of water management will have substantial and widespread impact through the West.

● *Groundwater is becoming increasingly contaminated.* About half the population depends on unseen groundwater resources for drinking water. Hundreds of thousands of Americans at one time or another have had to switch to bottled water. Experience has demonstrated that it is usually far less costly and far less complicated to protect groundwater from contamination than to clean up polluted supplies later. Effective protection will require modifying and controlling widespread, often diffuse activities, such as use of septic tanks and pesticide, herbicide, and fertilizer applications.

• *The review and reregistration of approximately 600 basic pesticide ingredients in use in this country are far behind where they should be.* Although this process was mandated in legislation in 1972, EPA estimates that it may take well into the next century to complete the job. In the meantime, environmental experts cite potential threats from pesticides as among the most important risks to public health and the environment. EPA itself has courageously identified these risks as far more serious than those posed by many problems on which EPA spends far more effort and money.

• *After eight years and $1.5 billion, the Superfund program to clean up toxic waste dumps has yielded disappointing results.* Distressingly little is known about how many toxic waste sites there are and how serious a risk each poses. No one has yet satisfactorily determined the standards to which sites should be cleaned, and the prospects of having agreement on these standards in any reasonable time are dim. Too few scientifically and legally defensible standards exist for troublesome chemicals. No one can yet say with authority that waste sites once sealed will remain so permanently.

With these fundamental questions unresolved, the task confronting EPA has been daunting, helping to explain why only 13 official cleanups had been achieved by mid-1987. Some cleanup efforts have begun at half the sites included on the National Priority List, and the agency has undertaken 1,000 emergency actions where the sites posed imminent threats. But the Superfund program may well result in many billions of dollars being spent with little net reduction in risk to public health and the environment. Hazardous waste cleanup, to which the nation has assigned a high priority backed by a $9 billion program, is hobbled by excessively onerous processes for fixing liability and mixing federal and private funding.

Superfund affords the best example of the lasting legacy of congressional mistrust of EPA, dating to the early years of the Reagan administration. The most promising, cheapest, and quickest solution to the cleanup of hazardous waste dumps is to obtain voluntary settlements. EPA has the legal authority and funds to foster such settlements, but Congress, which is highly suspicious of a repetition of sweetheart deals between regulators and industries, has created a climate hostile to settlements. As a result, there is little prospect the law's goals or timetables will be met or that its funding will stretch as far as had been hoped.

• *Degradation of wildlife refuges and national parklands continues.* More and more, the country's wildlife refuges and national parklands are threatened by energy, commercial, and other developments outside their borders. These developments not only diminish the visual and recreational amenities that parks and refuges provide to many millions of visitors each year, but also threaten the very survival of wildlife and undermine other natural and cultural resources these parks and refuges were established, at least in part, to protect. It often comes as a shock even to frequent visitors to the national parks to learn that some of the wildlife they are accustomed to seeing may soon no longer be there, as development in adjacent areas closes off vital habitat. Some major species, including the gray wolf and the mountain sheep, are likely to disappear from such parks as Yellowstone and Zion. Over the next century or two, Yosemite may lose between eight and fifteen of the twenty mammalian species that exist there now; Bryce Canyon may lose all its current large mammalian species. Climate change could wreak further havoc on parkland wildlife as changes in vegetation, temperature, and water availability alter habitat, even as surrounding development closes off potential escape routes.

• *Soil erosion continues at rates that are unacceptably high in the long term if the nation's farmlands are to continue supporting high levels of agricultural production.* An estimated 106 million acres (or 25 percent of U.S. cropland) exceed the average tolerable erosion levels each year. One especially costly dimension of this problem is the variety of problems caused by eroding soils when they leave the farm: for example, waterways polluted by pesticides and fertilizers carried off by erosion; reservoirs and harbors that silt up faster than predicted; recreational opportunities and wildlife habitat that are lost. The Conservation Foundation has estimated these losses off the farm (that is, not counting on-farm damages) at $6 billion per year.

• *Long after their values for flood control and fish and wildlife enhancement have been established, wetlands continue to be lost at a rapid rate.* With 50 percent of the nation's original endowment of wetlands now gone, draining, flooding, filling, cultivation, and development continue to destroy an estimated 300,000 to 500,000 acres per year. Some 80 percent of these losses have been attributable to agriculture. At the same time, many of the approaches used to manage wetlands are burdensome, costly, and inefficient. New programs are urgently needed to protect and manage the nation's wetlands more effectively.

• *After decades of environmental action by federal, state, and local governments and by citizen groups across the country, degradation of the American landscape is proceeding unchecked.* A steady, perceptible degradation of the countryside from urban sprawl and haphazard development continues to erode the distinctive qualities that differentiate one place from another. No national plan, no federal agency, can orchestrate protection of what Americans value about their communities; redefining the process and standards by which we build our landscape is ultimately a matter for local action. More than any other issue the poor quality of urban development—even in areas such as Florida, Colorado, and California—stands in stark contrast to the nation's lofty environmental aspirations. More than twenty years after the White House Conference on Natural Beauty, it is difficult to point to an example of urbanization in which the environment has not been degraded.

These examples hardly exhaust the list of problems. Carbon dioxide buildup in the atmosphere and the projected, accompanying climate change; depletion of the protective ozone shield, the loss of biological diversity; waste of energy; disposal of nuclear wastes; pollution at public facilities; loss of historic structures; threats to continuing productivity of national forests and rangeland; controversy surrounding the use of Alaska's abundant natural resources and the designation of wilderness areas there and in the continental United States; industrial and chemical accidents—

these are but a few of the dozens of issues that require urgent attention. For many of the long-standing environmental issues, there is at least a track record of programmatic successes and failures on which to build. Yet, in virtually every instance, the easy steps have been taken, the obvious solutions applied. The difficulty of making further progress on these issues, as well as on problems more recently recognized, is compounded by a number of factors.

• *The nation's budget situation provides a major source of conflict.* Competition among federal activities, including environmental programs, for a share of the U.S. budget is fierce. States and localities are little better off, for they have been asked to pick up many programs that the federal government has pared in efforts to reduce budget deficits.

• *The process by which policies are set and decisions made leaves much to be desired.* Better information is a requisite for better decisions. Yet the degree of uncertainty surrounding the data on which environmental decisions are based is often frightening. For example, many of the air quality models used to support regulatory decisions have enormous margins of error. Equally lacking is information about how well programs work; compliance statistics are notoriously incomplete, and monitoring of program implementation is problematic at best. Little evaluation has been done, for example, on the municipal sewage-treatment program to determine how much actual improvement in water quality has been brought by the billions of dollars the nation has invested. In addition, decisions are still too often made in the confrontational manner that has polarized environmental decision making in the past.

• *Many issues cut across the boundaries of traditional programs, thus requiring herculean efforts at integration and cooperation among multiple offices, agencies, and levels of government whose activities, as often as not, are competitive or adversarial.* Each of dozens of regulatory programs—for underground storage tanks, old hazardous waste sites, new hazardous waste sites, indoor air pollution, outdoor air pollution, workplace air pollution, and so on and on—concentrates only on its narrowly defined mission, generally ignoring often critical ecological interrelationships. Several different permitting systems exist at federal and state levels to regulate activities in wetlands, for example; as part of these programs, agencies use at least one-half dozen different definitions of wetlands, making compliance difficult and helping create interagency confusion and battles.

• *Some of today's problems are less visible, less tangible, and thus more difficult to mobilize for than those with which the country has been grappling over the past two decades.* Public outrage in this country over foul, dirty water or over brown-colored smog—problems people could readily see—helped marshal a constituency for environmental cleanup. The 1972 clean water legislation set a goal of "swimmable and fishable waters," a powerful image capable of capturing public attention and motivating action. People could see results. But no one can see acid rain or carbon dioxide or the ozone layer or indoor air pollutants or groundwater. Though the effects of these recently identified problems may be felt all the same, experts with access to sophisticated equipment and computer programs are increasingly needed to identify environmental problems and convince officials and the public of their scope and consequences.

• *The sources of many environmental problems are becoming far more diffuse.* Basic U.S. pollution control laws have relied on the states for their implementation and are premised on the assumption that damage caused by pollution occurs primarily in the state in which the pollution arises. Two decades ago, consensus for action in a community could develop over nailing a specific culprit—take action—and the problem would go away. Technology could reduce a power plant's emissions. A manufacturing plant could be required to treat its effluents. If pollution sources were numerous, they nonetheless were discernible, and strategies or controls for correcting the problems were available even if results fell short of expectations.

But it is much harder to target those who are responsible for the environmental problems now being recognized. Few strategies can be easily devised, few controls readily applied. No single culprit is causing the buildup of carbon dioxide and other gases in the atmosphere; countless individuals and economic activities, highly decentralized, are responsible. Similarly, depletion of ozone in the upper atmosphere can be laid at no one's doorstep in particular. Responsibility is shared widely. Tropical forests are falling, endangering wildlife, not only because of ill-conceived development projects but even more because of the activities of countless subsistence farmers eking out their living from the forests. Radon, a health hazard only recently recognized as such in this country, occurs naturally. Human activities are not to blame for the basic problem, though decisions by innumerable local officials over the years unknowingly have allowed homes to be sited in places of high exposure; the drive to create more energy efficient buildings has exacerbated the situation.

• *Some of the problems now confronting the country are likely to cause environmental and economic damage on a global scale.* Among these are climate warming; the threat of depletion of the earth's ozone layer, with its potential for increasing the incidence of skin cancer; the rampant loss of tropical forests and other highly productive ecosystems; and inadequate coordination and control of agricultural, chemical, and other goods traded in international commerce. No one country can solve these problems on its own. Different cultures and languages can turn the simplest transactions into complex undertakings. Above all, the weakness of international institutions, arising from jealously guarded national sovereignty, means that implementation would remain extremely difficult at best, even if nations could agree on cooperative measures.

When a new presidential administration takes office in January 1989, it will hold the potential for a fresh start and new leadership. It would be well advised to avoid the mistakes of ideology, single-mindedness, and confrontation that in the past have interfered with constructive development and implementation of sound environmental policy. This will be true for the broad range of environmental issues facing this country— pollution control, land management, wildlife protection, and so forth. The new administration should set a high priority on constructing an active, effective partnership among diverse interests. Numerous successful efforts at collaborative problem solving among business representatives, leaders of public interest groups, and officials at all levels of government show that such an approach is promising.

Unquestionably, the new administration would do well to begin the difficult process of overhauling the nation's current approach to pollution control. The administration will find no readily mobilized constituency for such a massive restructuring; too many interested parties—in Congress, conservation groups, industry, the legal profession, and elsewhere—have a stake in the current, heavily fragmented system. Federal legislative responsibility for the environment is split up among Congress's many committees, with little coordination of efforts.

Many of today's environmental problems defy traditional categorization. Acid rain, global climate change, groundwater pollution, toxic substances, hazardous waste—none of these problems fit into the way pollution control programs have been conceived in the past.

One important step forward in overcoming this fragmented approach is already being taken by EPA. The agency is devoting increased attention to risk assessment, the setting of priorities based on analysis of how many people are affected how seriously by multiple sources of exposure to contaminants. Setting priorities, however, requires consideration not only of environmental effects but also of factors such as feasibility and costs of curbing problems, and this could generate substantial conflict as various interests in Congress, EPA, and the public argue over what should be the priorities.

Regardless of the controversy, however, public consideration of a new approach to environmental protection is essential. Merely patching up current policies and institutions, helpful as it might be in the short term, simply will not be adequate for the country in the long term. Fundamental changes in concepts, in laws, and in the organizational structure of legislative and executive branch activities are essential if further progress is to be made on long-standing environmental issues and newly recognized ones alike.

The Conservation Foundation in the 1980s, even while it has pointed to the nation's unmet environmental needs, has repeatedly emphasized the indisputable achievements of the nation's environmental reforms of the 1970s. Those achievements deserve recognition, for they vindicate significant efforts and large financial sacrifices. Yet with time the victories over conventional air and water pollutants loom less large in the unfolding picture of both accumulated toxic wastes and emerging new problems that threaten groundwater, tropical and temperate forests, the upper atmosphere, and life on earth.

In the face of budget deficits, trade barriers, arms control, and other prominent national issues, even ensuring that the environment is on the agenda may not be simple. But the task is no less important for its difficulty. Achieving a prosperous and sustainable economy built on healthy and productive natural systems is critical as the nation prepares to move into the 1990s.

6

Encounter with the Reagan "Revolution"

Stewart Udall

"THE (REAGAN) ADMINISTRATION has no commitment to the environment and no environmental policy."
—Anne M. Burford, President Reagan's first EPA administrator (1985)

After nearly two decades of gains under five presidents, environmentalists were stunned in 1981 when the new secretary of the interior, James Watt, characterized their goals as "extremist" and announced that the Reagan administration intended to "reverse twenty-five years of bad resource management." But Watt's picadorish pronouncements were only the beginning. When all the ramifications of Reaganomics were spelled out, it was clear that the administration believed it was wrong to promote specific environmental goals.

Ronald Reagan's ideology about the evils of big government impelled him to reject the national consensus about resource stewardship that began with President Theodore Roosevelt. What made Reagan the first overtly anticonservation president of this century was the belief he shared with President Calvin Coolidge, that "the business of government is business."

Reagan wore blinders that centered his attention on a short list of core concepts. His predominant precept was the conviction that if stifling intrusions on the federal government were minimized (i.e., if, in his pet phrase, the government "got off the back" of American business), the operation of free markets would propel the nation into a new era of affluence and power. It was this concept that encouraged the Reaganites to dismiss national stewardship as an outdated idea. Planning and action to manage the nation's resources to secure environmental benefits for future generations would simply be superfluous once their free-market cure-all was functioning.

Of course, environmentalism was not the only target of the new president's ideology. The dogma that has dominated the Reagan years is distinguished by these axioms:

- a conviction that the federal government (with the notable exception of the Pentagon) was a bloated bureaucracy that had to be shrunk;
- a tenet that many environmental laws and regulations were choking industrial growth;
- a thesis that dynamism animated by unregulated markets, not initiatives by governments, would ameliorate—and ultimately resolve—the nation's social and environmental ills;
- a credo that the states were better situated than the federal government to manage the nation's resources and help citizens cope with social problems;
- a belief that the operation of free markets would solve the nation's energy predicament.

This ideological reorientation accounts for President Reagan's zeal for deregulation and his assiduous efforts to shrivel the functions of most nonmilitary agencies. It also illuminates the logic behind the Reaganites' knee-jerk assumption that every environmental gain would burden the economy with a corresponding economic loss for business. And it explains why Reagan's first director of the Environmental Protection Agency, Anne

Burford, began her incumbency with the assumption that EPA's main mission was not to protect the environment, but to cushion the impact of "unreasonable" laws on the industries it was her duty to regulate.

The same assumption guided Interior Secretary Watt's drive to "restore resource management to the people." Watt's rapport with Ronald Reagan emboldened him to challenge the "power and self-righteousness of Big Environmentalism" by flouting existing environmental laws as he carried out his plans to increase drilling, mining, and logging on the public lands. And to dramatize his disdain for environmentalists' concerns about the potential impacts of ill-planned offshore drilling, Watt ordered an unprecedented 1-billion-acre sale of oil leases on the continental shelf.

Using national security needs as his cloak, Watt delegated authority to the states to police strip mining and launched an effort to inventory all minable minerals in areas that were being considered as additions to the nation's wilderness estate. And he served notice on the American people that even popular conservation programs had to be sacrificed when he announced the phaseout of the Land and Conservation Fund, which for two decades had financed an unprecedented expansion of federal and state parkland, wildlife, and outdoor recreation programs.

The Reagan agenda was given momentum also by administrative changes designed to alter the "anti-industry" policies of previous administrations. In some instances environmental regulations were relaxed, and in others tacit decisions were made that laws on the books would not be enforced. In the case of the Superfund statute enacted by Congress in 1980, EPA officials spent so much time cozying up to the chemical companies that for several years nothing was done to clean up the dumps of toxic chemicals that were threatening the health of many communities. And despite the safety fiasco at Three Mile Island, the Nuclear Regulatory Commission was packed with proponents of atomic power who were so intent on accelerating the licensing of nuclear power plants—and on minimizing the implications of nuclear "accidents"—that they showed little interest in performing their duties as regulators of radiation safety.

But it was in the area of energy economics that Ronald Reagan's nonstewardship stance did the greatest harm to the nation's future. Leadership by presidents and by Congress in the 1970s made it possible for the country to pass rigorous energy-conservation laws and conduct successful campaigns to reduce dependence on imported oil. These efforts encouraged the American people and U.S. industries to make decisions about homes, autos, public transportation, and industrial plants that saved hundreds of billions of dollars and put the nation on a path of energy efficiency.

During the Reagan years, the government has abdicated leadership of this effort. Federal funding for energy research has been drastically cut, the rules that compel Detroit to produce more efficient autos have been relaxed, and the Reaganites have not only sought to dismantle the energy-conservation programs they inherited, but have systematically opposed new legislation to foster energy efficiency. Reagan's decision that the open market—and not federal leadership expressed through energy-conservation laws and programs—should determine the nation's energy policy is a judgment that is now setting the stage for an economic disaster in the 1990s as our commitment to end energy waste slackens and our dependence on imported oil increases each day.

President Reagan's stance that the nation's decisions about resources should be made in the marketplace made it difficult for conservationists to develop a response to the new administration. The president could have provoked a meaningful national debate had he confronted Congress with legislation to repeal or revise the laws his administration disliked. He might, for example, have urged that the reach of the Wilderness Act be cut back, or that the authority of the Environmental Protection Agency be circumscribed. Or he could have submitted bills to "return" most of the nation's public land to the states or recommended the outright repeal of statutes such as the National Environmental Policy Act and the Endangered Species Act that the White House viewed as obstructive or absurd.

Such blunt proposals would have precipitated a meaningful national debate, but the administration's tactical decision to avoid such showdowns meant that many vital issues would be contested on two battlegrounds. The first was the budgetary arena where the administration attempted, with considerable success, to achieve a de facto repeal of "bad" laws by using the president's budget to strangle resource programs it opposed. The second was the courts, where lawyers representing the Environmental Defense Fund (EDF), NRDC, and the Sierra Club Legal Defense Fund filed lawsuits to force officials of recalcitrant federal agencies to obey and enforce existing laws. After 1981, some of the best talent of the environmental movement was used to persuade federal judges to compel Secretary Watt and his successors, and other Reagan appointees, to faithfully execute important laws.

The environmental movement's overall response to the Reagan revolution demonstrated resilience and staying power. The alarms sounded by Reaganites swelled memberships of the national organizations from 4 million in 1981 to roughly 7 million in 1988. Adversity encouraged groups to form new alliances at the national level and fostered the formation of aggressive local coalitions to counter anticonservation initiatives emanating from Washington.

Moreover, while the Reaganites were dragging their feet on the protection of endangered species, the "quiet man" of the environmental movement, The Nature Conservancy, filled the gap by becoming the nation's prime acquirer and protector of the habitats of endangered plants and animals. And problem-solving initiatives of versatile organizations such as EDF and NRDC enabled them to outperform the federal government in helping embattled cities, states, and industries develop innovative solutions to regional water, energy, and other resource problems.

It is ironic that an administration that disavowed the concept that each generation owes a moral duty to future generations actually galvanized Americans to higher levels of commitment and action to save Earth's ecosystems. It is clear that on environmental issues, Ronald Reagan rowed against the American mainstream for eight years. The "great communicator" was unable to persuade Congress to repeal any important law he disliked. He has the shameful distinction of being one of only two or three presidents in this century who served his term without proposing any major initiative to further the cause of conservation. And all available evidence suggests that he did little to influence American attitudes on ecological matters. Indeed, a 1987 public-opinion poll revealed that environmental quality "has again become an consensual issue . . . as it was in the early 1970s."

I am convinced historians will not only indict the Reagan administration for its lack of vision concerning resources and its abdication of the traditional U.S. role of leadership in global environmental matters, but will conclude that Ronald Reagan's negative legacy is a gargantuan debt that restricts the action options of his successors and of the American people.

7

Forecast for Disaster

Robert H. Boyle

AND NOW THE NEWS for July 4, 2030:
  • *The second hurricane of the year has struck the East Coast. The 15-foot seawalls built to protect Baltimore, Philadelphia, New York, and Boston held against 12-foot tides, but a 25-foot storm surge swept over the eastern tip of Long Island, drowning 260 residents who had refused to leave their homes despite a federal evacuation order. The toll of dead on Martha's Vineyard, Nantucket, and Cape Cod is estimated at 50. The 310 fatalities are still far fewer than the 5,600 people who drowned in last month's hurricane in South Florida.*
  • *Twenty-two inches of rain from the hurricane flooded Washington, D.C., breaking the heat wave that had gripped the city for 62 straight days of 90°-plus temperatures. This fell short of the record set eight years ago when 72 consecutive 90-plus days caused the move of the nation's capital to the cooler environs of Marquette, Michigan.*

• *In Sepulveda, California, neighbors hammered an elderly widow to death when they learned she had been secretly watering a pot of geraniums. A footnote to this grim story: The woman's husband had died of thirst during the California drought of 1998.*

• *Food riots broke out in France, where vineyards and farmlands have turned arid amid the rising temperatures.*

• *Dust bowl conditions continue in the Plains States of the United States, but orange production is up in Saskatchewan. In eastern Siberia the outlook for a good cotton harvest is promising.*

• *In Stowe, Vermont, botanists announced the death of the last red spruce. The species' demise is blamed on a combination of stresses—acid rain, global warming, and ultraviolet radiation.*

• *In baseball, the Anchorage Braves beat the New York Mets 5–3. In Los Angeles, the Dodgers' game against the Calgary Giants, scheduled for the usual 5:30 a.m. start, was postponed because of dust storms.*

• *And now the weather. After leaving a swath of destruction in its wake along the East Coast, Hurricane Bruce is expected to move out to sea during the night. In the Midwest, Southwest, and West, conditions remain normal—searing heat, drought, and dangerous levels of ultraviolet radiation.*

If this reads like a newscast from *Saturday Night Live*, it isn't. This report has been extrapolated from carefully considered forecasts for our planet by a wide variety of scientists as we spin toward the twenty-first century.

Pollutants are saturating our atmosphere. Acid rain, which already has had a devastating impact on parts of eastern North America, central Europe, and southern Scandinavia, is one manifestation of this pollution, but its effects tend to be regional. Two similar and interrelated pollutant threats loom even larger, and they may soon affect life on a global scale. Both have the potential of wreaking catastrophic change on the earth's climate—and on life.

The first of these threats is the pollution caused by the release of chlorofluorocarbons into the atmosphere. These manufactured chemical compounds—more commonly called CFCs—are used as refrigerants and coolants and in the manufacture of everything from pillows to polystyrene boxes for fast food. Ever since their invention not quite sixty years ago,

CFCs have been rising into the stratosphere. When they hit the protective cover known as the ozone layer—10 to 20 miles up—they raise hell because their chlorine component devours the molecules that form the thin ozone shell. As that layer is depleted, stronger and stronger doses of ultraviolet (UV) radiation from the sun are able to penetrate to the earth's surface. Skin diseases and plant destruction are only the beginning of the troubles that excessive UV radiation can cause.

The other major threat is caused by the continuing buildup of carbon dioxide, nitrous oxide, and trace gases, including CFCs, in the atmosphere. In the 150 or so years since the industrial revolution, man's activities have enormously increased the atmospheric concentrations of these gases. The rapidly expanding use of fossil fuels and the vast destruction of the earth's forests have combined to create a great effusion of these so-called greenhouse gases. They are given that name because when they rise into the atmosphere, they form a kind of blanket in the sky that lets in solar heat but prevents heat from escaping the earth's atmosphere—much like a giant greenhouse. The resulting rise in air temperature could create havoc.

This is not the stuff of the far-off future, either. To the alarm of many scientists, a seasonal hole has begun to appear in the ozone layer above the Antarctic. When a significant drop in the ozone level was first recorded in 1978, the scientists who made the observations didn't pay much attention to their own data because no one had foreseen the possibility of such a thing. Unlike the ozone hole, the greenhouse effect was something scientists had anticipated, but it is developing faster than expected. In fact, Dr. James Hansen of the NASA Goddard Institute for Space Studies in New York flatly says that within 10 to 15 years the earth will be warmer than it has been in 100,000 years. Clearly, changes are under way. Whether they will be moderate or catastrophic depends on how humans respond.

CFCs were invented in 1930 by the late Thomas Midgley, who left another dubious legacy, tetraethyl lead for gasoline. Midgley came up with CFCs when the Frigidaire division of General Motors asked him to find a safe replacement for the toxic ammonia then used in refrigerators. When Midgley's discovery was placed on the market, it was quickly hailed as a miracle compound, and similar substances were created and adapted for a wide variety of industrial applications. Besides serving as refrigerants, CFCs came to be used as foaming agents, blowing and cleaning agents, and propellants in aerosol sprays. Now they are literally all over the place. The major industry trade group, called the Alliance for Responsible CFC Policy, notes that chlorofluorocarbon refrigerants are used to cool 75 percent of the food consumed in the United States, as well as for air conditioning in residential, industrial, and automotive applications. They are used as solvents to clean microchips and printed circuit boards and are mixed with ethylene oxide to produce a nonflammable gas that sterilizes hospital and pharmaceutical equipment. The same gas blend is also used as a fumigant and pesticide in granaries, warehouses, and ships' cargo holds. CFCs are used extensively in the production of plastic foams that insulate buildings, pipelines, storage tanks, railroad cars, and trucks, likewise the foam in pillows, cushions, mattresses, and the padded dashboards of cars; in egg cartons and in containers and cups for hot foods and beverages. When CFCs escape from discarded air conditioners and refrigerators, or when a bulldozer in the town dump crunches a discarded foam pillow or old mattress, the substances containing the CFCs are broken down, and the chlorofluorocarbons enter the atmosphere to do their dirty work in the ozone layer.

The most outspoken scientist on ozone depletion is a chemist named Sherwood Rowland. After receiving a Ph.D. at the University of Chicago, Rowland, now 60, earned an international reputation in radiation chemistry. In 1964 he became the chairman of the chemistry department at the University of California at Irvine. When he attended an Atomic Energy Commission meeting on atmospheric research in Fort Lauderdale in 1972, he was casting about for new fields to explore. At the AEC conference Rowland learned that James Lovelock, the unorthodox British scientist best known today as the father of the Gaia hypothesis—that all life on earth should be considered a single living entity—was going to report in the journal *Nature* that he had measured CFC levels in the lower atmosphere. In his paper, Lovelock suggested that CFCs might be used as atmospheric tracers, but he pronounced them "no conceivable hazard." Rowland was intrigued by the report; he had done research on fluorine, which is one of the components of chlorofluorocarbons, as well as in photochemistry (the action of light on chemicals), and he thought it might be interesting to study the eventual fate of CFCs in the atmosphere.

When Rowland began his investigation at UC Irvine in October 1973, the annual production of CFCs in the United States was on the order of 850 million pounds. DuPont, which sold them under the trade name Freon, was the major domestic manufacturer. Rowland did his initial research with Mario Molina, a postdoctoral student who had just received his Ph.D. from Berkeley. By December of that year the two scientists had completed their research, and in June 1974 they published a paper in *Nature*. The results of their research were startling, but as Rowland says, "There was no moment when I yelled 'Eureka!' I just came home one night and told my wife, 'The work is going very well, but it looks like the end of the world.'"

Briefly put, Rowland and Molina reported that CFCs were being added to the environment in steadily increasing amounts, that they aren't destroyed in the troposphere (the lower atmosphere), and that they survive for many decades, slowly drifting up into the stratosphere. Once CFCs reach the stratosphere, though, UV radiation decomposes them and releases chlorine atoms. This, in turn, triggers a catalytic chain reaction in which a single chlorine atom can destroy hundreds of thousands of molecules in the ozone layer before it eventually falls back to earth.

Ozone is constantly created by the action of sunlight on oxygen molecules, but over time chlorine atoms from relatively few decomposed CFCs can destroy more stratospheric ozone than the sun can create. The ozone layer is shifting and amorphous. It is thinnest and reaches its maximum altitude in the high stratosphere over the tropics, which is where most of the ozone is produced. The layer is at its lowest over the poles.

Rowland and Molina pointed out in their 1974 report that almost all the CFCs that had been released since the 1930s were still in the lower atmosphere, and thus the effect on the ozone layer could be expected to intensify in the future. Last May, Rowland told a joint hearing of the Senate Subcommittee on Environmental Pollution and the Senate Subcommittee on Hazardous Wastes and Toxic Substances that certain CFC compounds—notably CFC-11, CFC-12, and CFC-13—have lifetimes in the lower atmosphere that range from 75 to 120 years. "A 120-year average lifetime, without any intervening major changes in the atmosphere, means that . . . even without any further emission of [CFC-12]—and releases are occurring daily all over the world sufficient to average about 400 kilotons annually—appreciable concentrations . . . will survive in the atmosphere for the next several centuries."

But the publication of the Rowland-Molina report was just the beginning of the battle against CFCs. The Governing Council of the United Nations Environment Programme convened a panel of experts to examine the problem in 1977. The following year, Canada, Sweden, and the United States banned the use of CFCs in aerosol sprays (but only a few other countries have followed suit and CFCs from aerosol sprays still account for about 15 percent of the global total, according to the Environmental Defense Fund). In March 1985, after eight years of continued UN-sponsored meetings, the United States and twenty other countries signed what is now known as the Vienna Convention for the Protection of the Ozone Layer. The convention called for international cooperation in research and monitoring. It also provided for the adoption of international protocols to limit the emission of ozone-depleting substances, should such measures be necessary. Richard Benedick, a career diplomat who was the American deputy assistant secretary of state for environment health and natural resources, signed the document for the United States, calling it "a landmark event. It was the first time that the international community acted in concert on an environmental issue before there was substantial damage to the environment and health."

Two months later, in May 1985, *Nature* published alarming new information about CFCs. This paper was written by Dr. Joe Farman, an atmospheric scientist with the British Antarctic Survey, which had been routinely measuring the ozone layer above the Antarctic since 1957. He and others examined the data and saw that in recent years the ozone levels in September and October (the Antarctic spring) had fallen considerably.

The British measurements came from ground-based observations, and the wary Farman wondered if NASA satellites had recorded the phenomenon from space. At first it appeared that they had not. However, further checks of NASA computer data revealed that the hole in the ozone layer was apparent as early as October 1978—the first year in which such satellite comparisons could be made—and had reappeared each year at roughly the same time. The Farman paper suggested that the ozone drop might be tied to CFCs. But other scientists thought the unique weather dynamics above Antarctica were a more important factor. In August 1986, Dr. Susan Solomon, an atmospheric chemist with the National Oceanic and Atmospheric Administration, led a team of scientists to the Antarctic to study the hole. At its maximum, it was the size of the United States. The scientists also noticed that some ozone depletion extended as far north as Tierra del Fuego and Pantagonia. This past August four more teams traveled to Antarctica to make further observations. Although scientists are still going over their data, there now seems to be general agreement that the ozone hole is caused primarily by chlorine from CFCs.

Depletion of the ozone layer increases the amount of ultraviolet radiation reaching the earth, and the potential effects on human health are considerable. First, there's skin cancer. It is the most common form of cancer in this country, with an estimated 500,000 cases discovered each year. A study published by the Environmental Defense Fund projects that by 2025 there will be an additional 1.4 million incidences of skin cancer over the present rate if nothing is done to control ozone depletion.

Cataracts are another threat posed by elevated UV levels. So is alteration of the immune system. Research on the effects of UV radiation on the immune system has been done with mice as subjects. According to congressional testimony by Dr. Margaret L. Kripke, chairman of the department of immunology at the University of Texas, "There is considerable evidence that the UV rays damage a type of immune cell found in the skin, the Langerhans cell, and that this damage leads to activation of suppressor lymphocytes, instead of the appropriate immune response. Thus, although the initial damage is localized to the area of skin exposed to the UV radiation, the resulting immunological suppression is systemic, because the suppressor cells circulate throughout the body."

Not only humankind is at risk. Experiments with marine organisms have shown that UV radiation can damage animals in the marine food chain. The potential for damage to vegetation is also high. Dr. Alan Teramura, a professor of physiological ecology at the University of Maryland, reports that although some plants may adapt to UV radiation, many are adversely affected by increased levels. In tests, higher levels of UV radiation caused plant stunting, reduction in leaf area, and reduced physiological vigor—the latter rendering them more vulnerable to pests and disease.

In a six-year study of soybeans, UV radiation was increased to simulate a 25 percent reduction in the ozone layer: The result was a 20 to 25 percent loss in yields.

"Unlike drought or other geographically restricted stresses, increases in UV would affect all areas of the world simultaneously," Teramura says. "Even small reductions in crop yield on a global basis could lead to considerable economic consequences." Almost all knowledge of the effects of UV on plants comes from studies of cultivated crops, but these account for less than 10 percent of the world's vegetation. We have little or no information on the effects on the other 90 percent—the forests, grasslands, and shrub lands. In fact, there is much we don't now know about the extent of the damage that may be done by CFCs rising into the sky, because nothing like it has ever happened before. But when it comes to massive changes in climate, there are some precedents that may give us signs of what to expect.

Over the last 2,000 years, the earth has undergone two major changes in climate. The first was a warm period known to scientists as the medieval warm epoch; it occurred between the years 800 and 1250, when average global temperatures were about the same as they are now. Certain areas, however, were distinctly warmer. During that time barley and oats were grown in Iceland and vineyards flourished in England, where sea levels were gradually rising. In Belgium the rising sea made Bruges, now some 15 miles inland, a seaport.

Around 985, the Vikings began to colonize Greenland, which had been discovered by Eric the Red. But by the end of the thirteenth century Arctic sea ice had spread through Greenland's waters and had become such a navigational hazard that the colonies died out.

The medieval warm epoch was soon followed by the Little Ice Age, which lasted from about 1550 to 1850, during which the global climate was generally about 1°C (2°F) cooler than now. In India, the monsoons often failed to arrive, prompting the abandonment in 1588 of the great city of Fatehpur Sikri because of lack of water. The Thames froze over several times in the late 1500s. Year-round snow, now absent, covered the high mountains of Ethiopia. The vineyards of northern France died off.

Some scientists who have studied the earth's climatic cycles believe that around 1700, when the Little Ice Age began its gradual decline, the earth swung into a period of 1,000 years of natural warming. This forecast, however, does not take into account the effect of *unnatural* agents, such as the increasing concentrations of carbon dioxide, nitrous oxide, and other greenhouse gases in the atmosphere.

What's happening is this. Light from the sun passes through these transparent gases to the earth, where the shortwave radiation (light) becomes long-wave radiation (heat). The heat rises from the earth and ordinarily would escape into space. However, greenhouse gases absorb the long-wave radiation. Thus, the more these gases accumulate in the atmosphere, the more heat they absorb, and the warmer the earth becomes. In time, the planet will come to be like a greenhouse—or a car parked with its windows up on a sunny day.

The theory that increasing levels of carbon dioxide could cause this greenhouse effect was first advanced in 1896 by a Swedish physicist and chemist named Svante Arrhenius. However, the idea took on startling new significance in 1958 when Charles D. Keeling, a chemist and professor of oceanography at the Scripps Institution of Oceanography, began measuring atmospheric carbon dioxide on Mauna Loa in Hawaii. Since Keeling's measurements began, the concentration of the gas has increased every year. It jumped from 315 parts per million (ppm) in 1958 to 349 in 1987 — a 25 percent increase from the levels that are thought to have been present before the industrial age. The increase is attributable to a combination of the burning of fossil fuels and the destruction of forests, which serve as reservoirs of carbon. A forest stores about 100 tons of carbon per acre, and in the last forty years it is estimated that as much as half the world's forests have been destroyed. Given current emission levels, the atmospheric concentration of carbon dioxide is expected to reach about 420 ppm by the year 2030.

Two other greenhouse gases, CFCs and nitrous oxide, are double whammies: They are involved in the depletion of the ozone layer (in the case of nitrous oxide this is true only when the gas mixes in the atmosphere with CFCs or carbon dioxide) and they absorb heat. Measured in the range of parts per trillion, CFC concentrations might seem insignificant, but they are extraordinarily effective heat absorbers. One molecule of CFC-11 or CFC-12 can trap as much heat as 10,000 molecules of carbon dioxide. And CFC levels are increasing at the rate of 5 to 7 percent per year.

Ground-level ozone also qualifies as a greenhouse gas. It is formed by the action of sunlight on nitrogen oxide and hydrocarbon pollutants emitted primarily by cars and trucks. We call it smog. Ozone has a split personality. Stratospheric ozone protects life by shielding the earth from harmful UV radiation: Ground-level ozone is toxic. In the United States alone, according to a study made by the Environmental Defense Fund, ozone pollution is responsible for annual losses of as much as $2 billion in wheat, corn, soybeans, and cotton. Ozone produced on earth cannot be used to replenish the ozone layer in the stratosphere because it has a limited life span before combining into other chemical substances. Therefore it doesn't last long enough to accumulate in amounts significant enough to replace what's being lost in the stratosphere.

In the last 100 years, the global mean temperature has gone up by about 0.5°C. Even if all emissions of greenhouse gases were cut off today, past emissions already make another 0.5°C increase likely by 2050. According to computer model estimates done by Dr. Veerabhadran Ramanathan, an atmospheric scientist at the University of Chicago, the global average surface temperature could increase by a total of as much as 4.5°C in the next forty years, on the basis of current levels of greenhouse gas emissions. That would make the earth almost as hot as it was during the Cretaceous period, the age of the dinosaurs, 100 million years ago. Mind you, that is the global average. The greatest increase in temperatures will occur from the mid-latitudes to the poles, where wintertime averages could be 10°C higher than now.

Hansen, of NASA's Goddard Center, uses a climate model that predicts a temperature increase averaging 1° to 2°C in the United States by the middle of the twenty-first century. He also has created a computer model that predicts temperature increases for a number of U.S. cities. By around 2050—give or take a couple of decades because the role of the oceans is not yet predictable and could delay the warming effect—Washington, D.C., which according to Hansen's model has about 36 days a year when the temperature exceeds 90°F, will have 87 such days; Omaha, with 37 days over 90° now, will have 86; New York, with 15 now, will have 48; Chicago, with 16 now, will have 56; Denver, with 33, will have 86; Los Angeles, with 5, will have 27; Memphis, with 65, will have 145; Dallas, which has 100, will have 162. Hansen's model similarly shows an increase in 100°F days;

Washington goes from 1 a year to 12; Omaha from 3 to 21; New York from 0 to 4; Chicago from 0 to 6; Denver from 0 to 16; Los Angeles from 1 to 4; Memphis from 4 to 42; and Dallas from 19 to 78.

"Other discussions of the practical impacts of greenhouse warming have focused on possible indirect effects such as changes of sea level, storm frequency, and drought," Hansen says. "We believe that the temperature changes themselves will substantially modify the environment and have a major impact on the quality of life in some regions. . . . However, the greenhouse issue is not likely to receive the full attention it deserves until the global temperature rises above the level of the present natural climate variability. If our model is approximately correct, that time may be soon—within the next decade."

Dr. Wallace Broecker, a geochemist at the Lamont-Doherty Geological Observatory of Columbia University, thinks the situation may be even worse than indicated by models, with their supposition of a gradual warming over a considerable period of time. "The earth's climate doesn't respond in a smooth and gradual way," he says. "Rather, it responds in sharp jumps. These jumps appear to involve large-scale reorganizations of earth systems. If this reading of the natural record is correct, then we must consider the possibility that the major responses of the earth system to our greenhouse provocation will also occur in jumps whose timing and magnitude are uneven and unpredictable. Coping with this type of change is clearly a far more serious matter than coping with a gradual, steady warming."

These models are far from perfect—none of them was able to predict the ozone hole over the Antarctic, for example—but, for now, they're our best source of information about changes we can expect to see by the year 2050. The view is not pretty.

Climate modeling done by Dr. Syukuro Manabe, an atmospheric scientist at the National Oceanic and Atmospheric Administration Geophysical Fluid Dynamics Laboratory in Princeton, New Jersey, led him to testify before a congressional committee in 1985 that "winters in Siberia and Canada will be less severe. Because of the penetration of warm, moisture-rich air into the high latitudes, a doubling of atmospheric carbon dioxide or the equivalent might increase the rate of river run-off in northern Canada and Siberia by 20 to 40 percent. Our climate model also indicates that in response to the increased greenhouse gases, summer drought will become more frequent over the middle continental regions of North America and the Eurasian continent. For example, the model-produced summer drought is characterized by dry soil, reduced cloud cover, and higher surface temperature, which resemble the situation during the dust bowl of the 1930s."

A study by the National Academy of Sciences suggests that water volume in northern California rivers and in the Colorado River will decline by as much as 60 percent. This would leave much of the West without water. Southern California would run dry and be subjected to an increased incidence of fire, as would forests throughout much of the West and upper Midwest.

Within the past 100 years, tide gauges on the Atlantic Coast of the United States have documented a thirty-centimeter, or one-foot, rise in sea level. Globally, the average is about five inches. Models predict that the level will have risen by another foot in low-lying coastal regions of the country in 2030, and by as much as three feet in 2100. According to Dr. Steven P. Leatherman, director of the Laboratory for Coastal Research at the University of Maryland, at least part of the present sea-level rise on the East Coast is caused by the natural compacting and subsidence of coastal sediment. But at least 4.5 inches of the rise have been caused by the expansion of warmer ocean surface waters and the melting of mountain glaciers, triggered in part by the 0.5°C increase in global temperature registered during the last century.

"Sea-level rise will promote increased coastal erosion," Leatherman says. "Already approximately 80 percent of our sandy coastlines is eroding. . . . Artificial nourishment is being used to restore beaches, but the costs are high." According to one study that will soon be published, the cost of maintaining East and Gulf Coast beaches will run anywhere from $10 to $100 billion. A series of aerial photographs taken since 1938, for instance, shows that the Blackwater National Wildlife Refuge on the eastern shore of the Chesapeake Bay, one of the most important East Coast waterfowl sanctuaries, is in a state of disintegration because of rising sea levels. Human activity can hasten such destruction.

Some of the other threats posed by a one- to three-foot rise in sea level include increased salinity of drinking water; saline intrusion into river deltas and estuaries, which would imperil fisheries; the inundation of wetlands, cypress swamps, and adjacent lowlands; increased flooding in populated areas, which would necessitate the building of costly flood protection systems, such as seawalls; the disappearance of beaches all over the world.

Then there are these further dire possibilities:

• Studies by meteorologist Kerry Emanuel at MIT indicate that more severe hurricanes are likely because of warmer oceans. Such storms could increase in ferocity by as much as 60 percent over current maximums.

• Radical change in the Antarctic ice sheet could have severe consequences. Antarctica has 91 percent of the world's ice (only 1 percent is locked up in mountain glaciers). If the Antarctic ice sheets were to melt completely, the global sea level would rise 15 to 20 feet. No one expects that to happen. At currently projected rates, the greenhouse effect and global warming are not expected to have a major impact on the Antarctic ice sheet for several centuries. But no one predicted holes in the ozone layer, and as Dr. Stanley S. Jacobs, a senior staff associate at Lamont-Doherty, said in a recent article in *Oceanus* magazine: "Antarctica may be a wild card in the deck, but who can say the deck is not stacked, with Nature setting up the sting?"

Couple all the greenhouse effects with increased ultraviolet radiation, and we have written the prescription for disaster—ecological, economical, and political.

It is ludicrous to assume that we could rapidly adapt to such changes. "Infrastructures of society, such as water supplies, transportation networks, and land-use patterns have evolved over centuries in response to prevailing climate," says Dr. Gordon J. MacDonald, a former professor of geophysics at Dartmouth who's now vice president and chief scientist of the Mitre Corporation, a nonprofit research organization. "Significant changes in climate over decades will exert profound disruptive forces on the balance of infrastructures."

MacDonald is talking about infrastructures that are already in place. But corporations and governments throughout the world are now making big decisions about long-term projects that involve coastal development, massive land use, irrigation, hydroelectric power, oil exploration, natural gas, etc. Nearly all of these decisions are being based on the notion that the climate of the recent past will continue into the future. This is no longer a safe assumption. In October 1985 the World Meteorological

Organization, the International Council of Scientific Unions, and the United Nations Environment Programme convened a conference in Villach, Austria, at which more than eighty scientists from sixteen countries assessed the climatic changes that could be brought about by the accumulation of greenhouse gases. The scientists concluded that using the climate of the recent past to plan for the future "is no longer a good assumption since the increasing concentrations of greenhouse gases are expected to cause a significant warming of the global climate in the next century. It is a matter of urgency to refine estimates of the future climate conditions to improve these decisions."

Dr. Michael Oppenheimer, a former Harvard astrophysicist who is now senior atmospheric scientist with the Environmental Defense Fund, puts it this way: "We're flying blind into a highly uncertain future. These changes are going to affect every human being and every ecosystem on the face of the earth, and we only have a glimmer of what these changes will be. The atmosphere is supposed to do two things for us: maintain a constant chemical climate of oxygen, nitrogen, and water vapor, and help maintain the radiation balance—for example, by keeping out excess UV. The unthinkable is that we're distorting this atmospheric balance. We're shifting the chemical balance so that we have more poisons in the atmosphere—ozone and acid rain on ground level—while we're also changing the thermal climate of the earth through the greenhouse effect and—get this—simultaneously causing destruction of our primary filter of ultraviolet light. It's incredible. Talk about the national-debt crisis—we're piling up debts in the atmosphere, and the piper will want to be paid."

The fate of the earth rests on political decisions, which doesn't necessarily make it hopeless. Until recently, the Reagan administration has done little to deal with the crisis of atmospheric pollution. When the issue has been addressed, it has been largely at the prodding of individual legislators: in the Senate by Republicans John Chafee of Rhode Island, Robert Stafford of Vermont, and Dave Durenberger of Minnesota, and Democrats Max Baucus of Montana and George Mitchell of Maine, all members of the Environment and Public Works Committee.

Albert Gore, the Tennessee Democrat who's now a senator, led hearings on the greenhouse effect while he was in the House in 1981, and he's the first current presidential candidate to raise the issue. Indeed,

Gore's willingness to discuss this politically unpopular subject prompted columnist George Will to chide him for "a consuming interest in issues that are, in the eyes of the electorate, not even peripheral." But as Chafee says, "This is not a matter of Chicken Little telling us the sky is falling. The scientific evidence is telling us we have a problem, a serious problem."

Fortunately, it's still possible to ameliorate the damage. Here's what we must do:

• *Reduce production of CFCs by 95 percent worldwide within the next six to eight years.* Chafee and Baucus have introduced bills calling for such a reduction. Last winter Chafee told CFC manufacturers, "If the six- to eight-year phase-out in our bills is unrealistic, tell us how much time you need and show us how you will use that time. We are open to suggestions, but the burden is on you to justify a longer time frame.. . . Undoubtedly there will be testimony that we cannot ratchet down on production of CFCs too swiftly. It is well to recall that the ban on aerosols in the U.S. caused production of CFCs for aerosols to drop . . . to less than 25 million pounds . . . six years later. And our country survived. I am not convinced than American or any other producers have a constitutional right to continue to produce products that cause permanent harm to our world, to our citizens."

In September the United States and twenty-three other countries signed a treaty calling for a 50 percent cut in CFC production by mid-1999, but the new findings from the Antarctic demonstrate that the cut is neither big enough nor fast enough. "We've got to beat the clock," says Rafe Pomerance, a policy analyst who has been following the ozone problem for the World Resources Institute in Washington, D.C., for the past two years. "If the data from the Antarctic continues to build over the next few months, we may have to reconvene and strengthen the treaty."

• *Reduce dependence on fossil fuels.* "We should focus on incremental steps that limit our dependence on coal and oil," Oppenheimer says. "Let's focus on the doable. No. 1, conservation. The U.S. still uses twice as much energy per capita as the European nations. We're wasting money, we're wasting energy, and we're producing too much carbon dioxide because of our overdependence on fossil fuels."

Reliance on these fuels can also be reduced through greater use of nonpolluting alternative sources of energy. Solar power is a prime example, but the United States seems to have given up leadership in photovoltaic research, and the Japanese are now forging ahead. Photovoltaic technology promises to deliver energy at a reasonable price without producing carbon dioxide.

• *Halt deforestation.* "You have to do two things," says Dr. George M. Woodwell, former president of the Ecological Society of America and now director of the Woods Hole (Massachusetts) Research Center. "First, you have to stop deforestation around the world, not just in the tropics, and you have to do in on the basis of an international protocol. Second, you have to have an equally intensive and imaginative protocol that calls for reforestation so as to store one billion tons of carbon annually. A million square kilometers is 600 miles by 600 miles, and we will probably have to reforest on the order of 4 million square kilometers per year over good land to do the job."

• *Establish a national institute devoted to basic environmental research.* Says Oppenheimer: "We need a national commitment comparable to the Manhattan Project, not only so we can understand what the consequences of global change are for man, but so that we can be in the forefront of the development of alternative energy sources that will help limit this problem. I envision a multibillion dollar scientific effort. It's as important as national defense. It *is* the national defense. If we do nothing waiting for the atmosphere to change and for unpleasant consequences to occur, it will be too late for us to avoid disruptive and devastating changes."

• *Discontinue basic environmental research by or funded by EPA and the Department of Energy.* These agencies are unreliable because they are heavily influenced by political pressure. Last January, Broecker bluntly told the Senate Subcommittee on Environmental Protection, "I believe that most scientists would agree with me that the handling of research on greenhouse gases by DOE [the Department of Energy] and on acid rain by EPA has been a disaster."

Will the world act in time? As Rowland, who won eight varsity letters in basketball and baseball at Ohio Wesleyan and the University of Chicago, puts it, "The key thing about baseball is, there is always next year, another season. The question for the earth now is, will there be a next year?"

8

The Environment

Barry Commoner

IN THE PAST FEW DECADES, the United States has witnessed a series of remarkable popular movements, the most prominent being the movement for civil rights, for the equality of women, for peace, and for the environment. People who have participated in these efforts are prone to reflect after a time on what has been accomplished. In the case of the civil rights, women's, and peace movements, this is a subtle task, involving social processes that are not readily converted to "objective" numerical trends. To assess the impact of the environmental movement is easier, for the quality of the environment can be expressed in terms of generally unambiguous measurements: the changes, since the birth of the movement, in the amounts of harmful pollutants in the air or the water, or, for that matter, in our bodies; in the populations of fish or birds that have suffered environmental harm; in the efficacy of control measures.

The environmental movement is old enough now—its birth can be dated from the enthusiastic outburst of Earth Day, in April of 1970—to be held accountable for its successes and failures. Having made a serious claim on public attention and on the nation's resources, the movement's supporters cannot now evade the troublesome, potentially embarrassing question: What has been accomplished? Concern with the environment and efforts to improve it are today worldwide, but the United States is the place where the environmental movement first took hold, and where the earliest efforts were made, so it is a good place to look for answers. Since the early nineteen-seventies, the country has had basic laws that are intended to eliminate air and water pollution and to rid the environment of toxic chemicals and agricultural and urban wastes. National and state environmental agencies have been established; billions of dollars have been spent; powerful environmental lobbies have been created; local organizations have proliferated. Environmental issues have taken a permanent place in our political life.

In one respect, all this activity has clearly achieved an important success: We now know a good deal more about the state of the environment than we used to. In the last fifteen years, the United States has established monitoring systems that record the annual changes in the environmental quality, and these give us an indication of what has happened in the environment since the effort to improve it began. The amounts of different pollutants that are emitted into the environment each year can be estimated fairly accurately for the nation as a whole from technical data on the behavior of cars, power plants, factories, and farms, though such measurements do not reveal the levels of pollutants that people actually encounter, which may differ a great deal, depending on how far they live from the pollutants' source. Local concentrations of some common pollutants—for example, the amount of dust or sulfur dioxide per cubic meter of air—are measured by a network of air-sampling devices that are stationed at fixed points, chiefly in cities, but the resultant information is spotty and not readily translatable into an average national trend. Water taken from rivers, lakes, underground sources, and wells is also analyzed from time to time for chemical and biological pollutants, but, again, the results are necessarily discrete and localized. Finally, there are less comprehensive measurements that determine the amounts of certain pollutants, such as pesticides, that have been taken up by wildlife, especially fish and birds, and there are also a few corresponding analyses of pollutants carried in the bodies of the human population.

The nation's basic environmental law, the National Environmental Policy Act of 1969, assigns to the federal Council on Environmental Quality the task of reporting these data. Unfortunately, among the first of the Reagan administration's many cuts in domestic programs was a 62 percent reduction in the CEQ budget, which has diminished its reports, especially on water pollution. Nevertheless, by rounding out the CEQ reports with special ones produced by various other government agencies it is possible to piece together a picture of how the environment has fared in the last ten or fifteen years.

Information about the trends in air pollution is available from annual reports published by the Environmental Protection Agency for the years between 1975 and 1985. (Data earlier than 1975 tend to be unreliable, because measurements were not standardized.) The EPA reports describe changes in the emissions and local concentrations of the major airborne pollutants: particulates (dust), sulfur dioxide, lead, nitrogen oxides, volatile organic compounds, and ozone, a key ingredient of photochemical smog. One striking fact immediately emerges from the monitoring data: It is indeed possible to reduce the level of pollution sharply, *for between 1975 and 1985 total annual lead emissions decreased by 86 percent, and airborne concentrations of lead at national test sites have been correspondingly reduced.* Lead, a notoriously toxic metal, has been responsible for serious health effects, such as mental retardation, especially among children living in heavily polluted areas. People have benefited from the reduced emissions: *The average lead levels in the blood of Americans decreased by 37 percent between 1976 and 1980.*

The successful effort to reduce lead pollution only accentuates the failure to achieve a comparable reduction in the emissions of the other air pollutants, which, on the average, decreased by only 13.2 percent between 1975 and 1985. Of these pollutants, dust emissions have improved most—about 32 percent—although actual concentrations in the air at some 1,500 test sites have improved somewhat less. But the annual improvements came to a halt in 1982, and since then emission levels of particulates have increased: They rose by more than 4 percent between 1982 and 1985.

Sulfur dioxide is a particularly serious pollutant, for it diminishes the respiratory system's ability to deal with all other pollutants. It is also a major contributor to acid rain. Between 1975 and 1985, total sulfur-dioxide emissions declined by 19 percent, most of the change occurring between 1975 and 1981; since then, emissions have remained essentially constant. Average concentrations at national test sites improved somewhat more, in part because new power plants—a major source of sulfur dioxide—are being built outside urban areas, while most of the test sites are inside urban areas.

Carbon monoxide, a pollutant that causes respiratory problems, is produced chiefly by cars, trucks, and buses, and the effort to deal with it is based on control systems that reduce emissions from automobile exhausts. (The control devices also reduce waste-fuel emissions.) Total annual carbon-monoxide emissions decreased by 14 percent between 1975 and 1985, but between 1982 and 1985 they increased. In a number of cities, including New York, the levels of carbon monoxide still violate EPA standards.

Photochemical smog, a complex mixture, is created when nitrogen oxides emitted from automobile exhausts and power plants are converted by sunlight into highly reactive molecules and these combine with waste fuel to form ozone and other noxious chemicals. The total emissions of nitrogen oxides *increased* by 4 percent between 1975 and 1985. Ozone is not emitted as such but is formed in the air during smog reactions. Concentrations at national test sites decreased by 15 percent between 1975 and 1981, but have not changed since.

The noxious smog chemicals are responsible for serious health hazards; people with heart or respiratory problems are routinely warned to stay indoors during "smog alerts." This hazard is now more or less accepted as an unavoidable aspect of urban life. In some places, improvements in smog levels have been achieved by reducing traffic; yet smog continues to threaten health. For example, in Los Angeles, the worst-afflicted city, between 1973 and 1977 residents were subjected each year to at least 250 days in which smog was at levels classified as "unhealthful," with more than 125 of these classified as "very unhealthful." In most large American cities, residents are still exposed to unhealthful smog levels for fifty to 150 days each year.

One consequence of air pollution is acid rain. In keeping with the ecological rule that everything has to go somewhere, sulfur dioxide and nitrogen oxides, once they have been emitted into the air, are picked up by rain and snow and brought down to earth in the form of sulfates and nitrates. Both these substances increase acidity, and in recent years many lakes—especially in the northeastern United States and Canada, but also in Europe—have become more acid. In some of these lakes, serious biological changes have occurred, often involving the virtual elimination of fish populations. In cities, acid rain erodes buildings and monuments.

Because scrubbers and the increased use of low-sulfur coal have reduced sulfur-dioxide emissions at coal-burning power plants, while emissions of nitrogen oxides from power plants and automotive vehicles have increased, there has been a noticeable shift in the relative contributions of these pollutants to acid rain. Reports from Hubbard Brook, a research station in New Hampshire, where acidity problems have been studied for a long time, show that as the sulfate content of precipitation declined by 27 percent between 1964 and 1981 the nitrate content increased by 137 percent. Not much improvement can be expected in the acid-rain problem, given the negligible improvement in sulfur-dioxide emissions of nitrogen oxides.

The ecological processes that govern the quality of surface waters are more complex than the chiefly chemical events that govern air pollution. A basic reason for the pollution of surface waters—rivers and lakes—as well as inshore marine waters is the stress placed on the natural ecological cycles that, if they are kept in balance, maintain water quality. If inadequate sewage-treatment systems dump excessive organic matter into a river or a lake, the accompanying fecal bacteria threaten health. As the excess organic matter is broken down by aquatic microorganisms, they may consume so much oxygen that fish begin to die. If urban and industrial waste and runoff from agricultural areas increase nitrate and phosphate concentrations beyond the levels maintained by a balanced cycle, eutrophication occurs. Heavy algal blooms are formed and soon die, burdening the system with excessive organic matter and reducing oxygen content. In addition, high nitrate levels in drinking water may create health problems such as methemoglobinemia (a condition that reduces the oxygen-carrying capacity of the blood, especially in infants) and may contribute to the formation of carcinogens. Toxic chemicals only add to these harmful effects.

In the last decade, particular rivers and lakes here and there have been cleaned up by closing sources of pollution and building new sewage-treatment plants. Yet in that period, nationally, there has been little or no overall improvement in the levels of the five standard pollutants that determine water quality: fecal coliform bacteria, dissolved oxygen, nitrate, phosphorus, and suspended sediments. A recent survey of the trends in pollution levels between 1974 and 1981 at nearly 400 locations on major American rivers shows that there has been no improvement in water quality at more than four fifths of the tested sites. For example, the levels of fecal coliform bacteria decreased at only 15 percent of the river stations, and increased at 5 percent. At half the locations, the bacterial count was too high to permit swimming, according to the standard recommended by the National Technical Advisory Committee on Water Quality Criteria. Levels of dissolved oxygen, suspended sediments, and phosphorus improved at 13 to 17 percent of the locations, but deteriorated at 11 to 16 percent of them. The most striking change—for the worse—was in nitrate levels: Increases were observed at 30 percent of the test stations and decreases at only 7 percent. Agricultural use of nitrogen fertilizer is a main source of this pollutant; in rivers that drain cropland, the number of sampling stations that report rising nitrate levels is eight times the number reporting falling levels. Another major source is nitrogen oxides emitted into the air by vehicles and power plants and deposited in rain and snow as nitrate; this accounts for increased river nitrate levels in the Northeast, despite the relative scarcity of heavily fertilized acreage in that area. The survey also shows that there was a sharp increase in the occurrence of two toxic elements, arsenic and cadmium (a cause of lung and kidney damage), in American rivers between 1974 and 1981; but, as expected from the reduced automotive emissions, the occurrence of lead declined.

An overall assessment of the changes in these standard measures of water quality can be gained from the average trends. For the five standard pollutants, the frequency of improving trends averaged 13.2 percent, but the frequency of deteriorating trends averaged 14.7 percent; thus, at more than four fifths of the test sites, overall water quality deteriorated or remained the same. In sum, the regulations mandated by the Clean Water

Act, and more than a hundred billion dollars spent to meet them, have failed to improve water quality in most rivers. The relatively few locations that have improved are more than cancelled out by the locations that have deteriorated. Moreover, the occurrence of three serious pollutants—nitrate, arsenic, and cadmium—has increased considerably.

One of the chief symptoms of the environmental crisis in the early 1970s was eutrophication, especially in lakes. Lake Erie provided a dramatic example. The lake received an enormous burden of inadequately treated sewage and phosphate-rich detergents from the cities surrounding it, and chemical plants contributed mercury and other toxic materials as well. Rivers carried nitrate and eroded soil into the lake from heavily fertilized farms. By the 1960s and '70s, especially in its western regions and along the shoreline, Lake Erie was exhibiting the classic signs of eutrophication: heavy algal overgrowths, epidemics of asphyxiated fish, and sharply declining fish catches. Because of its notoriety as a "dying lake," Lake Erie has been intensively studied in the last decade, and the results have been evaluated by elaborate statistical techniques; the most detailed data are given in a recent lengthy EPA report. This report makes a telling comparison of pollution levels along various reaches of the western lakeshore, where eutrophication had been particularly troublesome. In 1972 and 1973, three of twenty-one shoreline regions were classified as entirely or partly eutrophic; in 1978 and 1978, eutrophication was more widespread, affecting twelve of twenty-one shoreline regions.

According to the EPA report, the rate of oxygen depletion in the central basin of Lake Erie increased by 15 percent between 1970 and 1980, following a trend that went back as far as 1930. The entry of phosphates from certain city sewage systems has declined, as a result of campaigns to reduce the use of phosphate-containing detergents—probably the most successful such effort, in Detroit, reduced phosphate concentrations in the Detroit River by nearly 70 percent between 1971 and 1981—but the acquisition of phosphates from all the rivers entering the lake was reduced much less; only four of twelve test sites showed any improvement. Overall, phosphates entering Lake Erie decreased by about a third between 1972 and 1982. The same EPA report also records the catch of commercial fish from 1920 to 1980. The once-valuable catch of herring and whitefish dropped off sharply after 1950, and by 1960 it was close to zero. The EPA report showed that the catch had not recovered between 1960 and 1980. In 1970, Lake Erie was widely mourned as a dying lake, and serious efforts were made to revive it. Yet nearly two decades after our environmental reawakening Lake Erie, despite some limited improvements, remains a flagrant example of environmental pollution. The condition of less famous lakes is just as bad. This is particularly true of lakes in heavily farmed areas, where nitrogen and phosphorus fertilizers leach from the soil into rivers and lakes. A 1982 survey found that of 107 lakes in Iowa all were eutrophic, and so were more than 80 percent of the lakes in Ohio and Pennsylvania. A national survey of changes in water quality between 1972 and 1982 showed almost no progress in lakes. Only 2.4 percent of the total lake acreage improved; 10.1 percent became more degraded; 62.1 percent were unchanged; no reports were available for the remainder.

About 50 percent of the population of the United States depends on underground sources—groundwater—for its drinking water. The United States Geological Survey and state agencies monitor the quality of groundwater by testing wells throughout the country. The results, based on readings from more than a hundred thousand wells, show that in the past twenty-five years these sources have become increasingly polluted by nitrates and toxic chemicals. Fertilizer was chiefly responsible for the rising nitrate levels. In Nebraska, a 1983 survey showed that 82 percent of the wells over the nitrate limit established by health authorities (ten milligrams per liter of nitrogen in the form of nitrate) were contaminated by fertilizer nitrogen. In California's Sacramento Valley, a very heavily cropped area, nitrate contamination of wells has been monitored for a long time. In the fifty-year period following 1912, the percentage of wells with excessive nitrates (defined as 5.5 milligrams per liter) approximately doubled. More recently, the percentage of wells with excessive nitrates doubled again— this time in only a four-year period, between 1974 and 1978. The major source of the nitrates is nitrogen fertilizer leaching from irrigation water. A similar trend has been observed in Iowa. In 1984, the Geological Survey summarized the situation: "Current trends suggest that nitrate accumulations in groundwater of the United States will continue to increase in the future."

Fifteen years ago, public-opinion polls on environmental issues showed that most people were worried about water and air pollution—especially smog. Now, even though these problems remain largely unsolved, polls show that as a public concern air and water pollution runs behind a new environmental threat—toxic chemicals. In the early 1970s, this problem was due largely to agricultural products—insecticides, herbicides, and fungicides. DDT and similar chlorinated insecticides were the most notorious examples. In 1972, the use of DDT and related insecticides was banned in the United States because they were shown to promote cancers and also to be a hazard to wildlife. One of the most noticeable effects of DDT was the decline in bird populations; DDT interferes with the biochemistry of reproduction, making eggs thin-shelled and thus easily destroyed before they hatch. Banning the use of DDT has been very effective. For example, between 1969 and 1975 the average DDT content of brown pelicans in South Carolina decreased by 77 percent, and by 1976 the number of fledglings more than tripled. People have benefited as well: Between 1970 and 1983, average DDT levels in body fat in the American population decreased by 79 percent. The banning of polychlorinated biphenyl, another notorious chemical pollutant—it increases the incidence of cancer and birth defects—has had a similar effect. Between 1970 and 1980, the body burden of PCB in freshwater fish decreased by 56 percent, and in birds (starlings) by 86 percent. In people, the percentage of Americans with relatively high levels of PCB (above three parts per million) in their fatty tissue decreased by about 75 percent.

Since 1950, however, the roster of serious chemical pollutants has steadily expanded. Hundreds of toxic chemicals, many of them carcinogenic, have been detected in water supplies, air, and food. For the first time in the three-and-a-half-billion-year history of life on this planet, living things are burdened with a host of manmade poisonous substances. According to a recent EPA survey, members of the general American population now carry several dozen manmade chemicals, many of them carcinogenic, in their body fat (and generally in the fat of mothers' milk as well). Toxic chemicals now seriously pollute important segments of the food chain. In the Great Lakes, for example, the toxic-chemical levels of salmon, trout, and walleye often exceed Food and Drug Administration standards for human consumption. Tumors are found on the fish with increasing frequency.

The total toxic-chemical problem is huge. The American petrochemical industry produces about 265 million metric tons of hazardous waste annually; toxic chemicals generally make up about 1 percent of this material, and the rest is made up of water and other nontoxic carriers. About a third of this waste is emitted, uncontrolled, into the environment. Moreover, most of the controlled, or "managed," waste is injected underground, sometimes into water systems, and thus becomes a long-term threat to the environment. Only about 1 percent of the industry's toxic waste is actually destroyed. The chemical industry has, largely unrestrained, become the major threat to environmental quality.

Environmental pollution from radioactive materials is in a class by itself; it originates from a single sector of production—the manipulation of nuclear energy for peaceful or military purposes. Radiation exposure due to the normal operation of the nuclear power industry appears to be quite small compared with exposure to natural sources of radioactivity such as radon and cosmic rays; it is less than .01 percent of the natural exposure. But this is a national average; near power plants, exposures are certainly higher.

There is no widely disseminated accounting of the radioactivity that enters the environment from minor nuclear-power-plant malfunctions. At most, there are only reports that an accidental emission of radioactive material has occurred—generally with no information about the actual radiation exposure but only the statement that "the amount of radiation released was harmless." (Such statements are wrong. Radiation, no matter how weak, always involves the risk of some harm, and the risk is proportional to the dose received. A very small exposure from a dental X-ray, say, has a correspondingly small risk, which is presumably worth taking in view of the expected benefit.) A partial measure of the radioactivity released by nuclear power plants is available from EPA studies of krypton-85, a radioactive gas uniquely associated with the operation of such plants. The average annual concentration of krypton-85 in the air increased by 80 percent between 1970 and 1983, thus increasing the environmental hazard.

Accidents that disrupt the containment structures protecting the environment from the huge amount of radioactive material in an operating nuclear reactor present a far larger hazard than is presented by typical malfunctions. Until 1979, this was an abstract concern, for some abstruse—and considerably disputed—statistical computations concluded that the problem of a serious accident might be as low as one in a million per year per reactor. The accident at the Three Mile Island nuclear power plant, in 1979, brought the abstract discussion down to earth. While the outcome of the accident was far less serious than it might have been—the reactor's core partially melted and extensively contaminated the interior of the reactor building and released some radioactivity into the environment—it suggested that serious mechanical failures were far more probable than the calculations had indicated.

In April of 1986, of course, a second failure occurred—at the Soviet Union's Chernobyl nuclear power plant—which led to a radioactive disaster. At least thirty-one people have died and more than 200 have suffered acute radiation sickness; estimates of future cancer deaths from the radioactive fallout over a wide area of Europe range above a hundred thousand; more than a hundred thousand people have been evacuated from their homes; some hundreds of square miles of agricultural land have become useless; radioactive fallout disrupted milk and vegetable production throughout most of Europe.

The hazardous consequence of one aspect of nuclear technology—fallout from test explosions of nuclear weapons in the atmosphere—has been considerably reduced by the straightforward procedure of simply stopping the process that creates it. As a result of the 1963 treaty signed by the United States and the Soviet Union, atmospheric tests have halted (except for a few conducted by China and France), and fallout radioactivity, in keeping with natural decay processes, has declined. For example, a national survey of milk showed that it contained 23.8 picocuries of strontium 90 per liter in 1964, 7.3 in 1970, and 2.0 in 1984.

Finally, we need to consider the American contribution to global environmental problems. One of these problems is the progressive depletion of ozone in the stratosphere, which is due largely to the increased use of chlorofluorocarbons (petrochemical products widely used in refrigerators and air conditioners). The stratospheric ozone layer protects the earth from solar ultraviolet radiation; if it is sufficiently depleted, people will be exposed to an intensity of radiation capable of significantly increasing the incidence of skin cancer. Recent studies in Antarctica indicate that the ozone layer is continuing to thin out. Despite an international agreement to reduce the production of chlorofluorocarbons, and some progress in the United States, little has been done thus far to stop this chemical assault on the stratosphere. Another problem is the progressive warming of the earth's surface because of the accumulation of carbon dioxide and other gases in the upper atmosphere. Sunlight falling on the earth is sooner or later converted into heat, which radiates outward, so the planet's temperature is determined by the balance between the sunlight falling on the earth and the heat leaving it. The concentration of carbon dioxide in the atmosphere governs the reradiation process, for the gas acts like an energy valve, allowing sunlight through but tending to hold heat back. Since 1850, the increased use of fuel that produces carbon dioxide when it is burned has gradually increased the atmospheric-carbon-dioxide concentration, and the earth's average temperature has steadily risen; an additional three or four degrees Fahrenheit in average temperature is expected by the middle of the twenty-first century. This may cause drastic changes in climate—for example, flooding of coastal areas (including many cities) and severe drought in North America, Europe, and Asia. In recent years, other gases, such as the chlorofluorocarbons, have added to the problem. As a result, the latest projections predict severe climatic changes by 2030.

This "greenhouse effect" (glass resembles carbon dioxide in its effect on sunlight and heat, a property that helps to keep a greenhouse warm in the winter) can be slowed down only if there is a sharp reduction in the use of combustible fuel, but nothing has been done thus far to deal with this problem, and it continues to carry the world toward a climatic catastrophe.

So much for the statistics on the changing state of the environment. What do they mean? Do they indicate that we can repel the assault on the environment by gradually reducing pollution at the present modest and uncertain rate? Can we accelerate the process by more stringent emission controls? Each year, the annual CEQ report has accompanied its environmental statistics with optimistic answers to such questions. The 1983 report, for example, began the chapter on environmental trends with this assertion: "As measured by traditional indices, such as the amount of pollutants in surface water and air, the overall quality of the environment in the

132

United States seems to be improving." This modest claim was inflated by President Reagan, who said shortly after the report appeared, "I am proud to report that the most recent studies of the Environmental Protection Agency show that we have made great progress in cleaning up the air and water" — a statement, hardly warranted by the facts, that one environmental leader described as "a big leap forward," because Reagan chose even to mention the environment.

Self-praise is perhaps natural from the President and the government agency that is responsible for improving the environment, but it is scarcely justified by the Administration's enforcement record. From 1973 to 1979, the annual number of civil enforcement actions by the EPA increased year by year, paralleling the modest decline in pollution levels, at least in the air. Then, after Reagan took office, budget cuts were imposed on the EPA most severely in those divisions that deal with enforcement. To no one's surprise, enforcement proceedings dropped off; between 1980 and 1981, the number of cases involving violations of environmental laws that the EPA referred to the Department of Justice decreased by 69 percent. More recently, with new EPA administrators, the caseload has increased, suggesting that the earlier decrease reflected Administration policy rather than some sudden decline in the number of environmental scofflaws.

These changes can be directly traced to policies established by President Reagan, who in transmitting a recent annual environmental report to Congress declared, "We have expanded innovative programs which allow industry the regulatory flexibility and economic incentives to clean up pollution." The term "regulatory flexibility" is the industrial lobbyist's well-known euphemism for relaxing the enforcement of regulations. And Reagan's message has obviously been received by American industry. Capital expenditures for new plant and equipment for pollution abatement by American industries, which had been at an average level of about $4.8 billion dollars per year (in constant 1972 dollars) between 1973 and 1980, have since declined, sinking to $3 billion in 1983. It is, of course, not surprising that Reagan's policies have been carried out by his environmental agencies and his industrial supporters. What was also foreseeable, and even more disturbing, was that these policies would affect the environment itself. So while the annual emissions of air pollutants gradually declined between 1975 and 1981, the year in which Reagan took office, they have since decreased much more slowly or not at all. For all the air pollutants other than lead, the average annual rate of decline in emissions was 1.52 percent between 1975 and 1981 but only 1.16 percent between 1981 and 1985. Indeed, since 1982 the long-term, if gradual, decline in air-pollutant emissions has come to a halt. Between 1982 and 1985, there was an average annual *increase* of 0.87 percent in these emissions; the emission of only one of the five pollutants (sulfur dioxide) declined. It has been argued that the halt in the rate of environmental improvement is due not so much to weaker enforcement and reduced abatement expenditures as to increased industrial activity during the period. However, industrial activity between 1975 and 1980 increased at about the same rate as it did between 1982 and 1985, but it was only in the earlier period that the major decline in air-pollution levels occurred. It appears that Reagan has left his mark not only on environmental policy but also on the environment itself—a mark that can be measured in tons of dust, sulfur dioxide, carbon monoxide, nitrogen oxides, and organic compounds that could otherwise have been removed from the nation's polluted air.

The lack of improvement in the acid-rain problem can also be traced to Reagan's policies. While his Administration has grudgingly acknowledged that there is an acid-rain problem, it has refused to accept the widely held view of its cause and has therefore done little to deal with it apart from a recent Administration promise to fund more research—an action that led a member of the Canadian Parliament to remark, "It won't reduce emissions by a single ounce."

Though environmental organizations are usually critical of Reagan's environmental stewardship, they nearly match his optimism about the course of environmental improvement. For example, the Conservation Foundation's 1982 report entitled "State of the Environment" claimed:

> The nation has made impressive progress in its attack on some conventional environmental problems. . . . Emission of most major air pollutants have continued to decline. . . . Available data show no similar progress toward water quality goals. Still, even to hold the *status quo* is an achievement in the face of significant economic and population growth since 1970.

It seems that at least some environmental activists are moderately happy about their accomplishments. But such optimism does not necessarily respond to the original thrust of the environmental movement, which envisioned not an environment that was a little less polluted than it was in 1970, or holding its own against an expanding economy, but an environment free of mindless assaults on ecological processes. By this standard, the question is whether the movement's goal can be reached by the present spotty, gradual, and now diminishing course of environmental improvement or whether some different course must be followed. To find the answer, it is helpful to distinguish between the reasons for the rather few environmental successes and the reasons for the more numerous partial or complete failures. We need to explain the striking fact that while the levels of most pollutants have declined only modestly, and others not at all, a few have been reduced quite sharply: lead, DDT and similar chlorinated pesticides, mercury in surface waters, radioactive fallout from nuclear-bomb tests, and, in some rivers, phosphates. Understanding these successes, which begin to fulfill the original aim of the environmental movement, may shed light on the reasons for the more common failures and suggest how they may be remedied. There are two general ways in which a pollutant may be prevented from entering the environment: Either the pollutant is eliminated from the activity that generates it, or a control device is added to trap or destroy the pollutant before it enters the environment.

The few real improvements have been achieved not by adding control devices or concealing pollutants (as by pumping hazardous chemical wastes into deep water-bearing strata) but simply by eliminating the pollutants. For example, the reason that there is now so much less strontium 90 in milk and in children's bones is that we and the Russians have had the simple wisdom to stop atmospheric testing of nuclear weapons, which produces it.

When the process that produces a pollutant is stopped — the banning of pesticides, the halt in atmospheric nuclear testing — there is considerable environmental improvement; if, instead, an effort is made to control the pollutant by recapturing or destroying it before it escapes into the environment, there is some improvement in environmental quality, but generally not much. In fact, such controls are ultimately self-defeating. To begin with, the effect of a control device is never complete. For example, unlike the absolute impact on lead pollution of the phasing out of leaded gasoline—after all, every ounce of lead not added to gasoline is also absent from the environment—the installation of an exhaust-control system cannot completely halt carbon monoxide pollution. Even when the system is in perfect working order, it does not trap all the carbon monoxide but only about 90 percent of it. Besides, the effectiveness of the catalyst rapidly declines with use, so any delay in replacing it means increased pollution. As long as control devices are not absolutely perfect—and none is—continued increase in the pollution-generating activity (traffic, for example) will gradually overwhelm the devices' effect on environmental quality. Finally, control devices cannot be used at all if the pollutants' points of entry into the environment are so numerous that they are impossible to identify. The classic example is the nitrates that leach from heavily fertilized soil into rivers and lakes along every inch of their banks.

All these reasons explain why those environmental pollutants—concentrations of carbon monoxide, sulfur dioxide, and dust, for example—that are now subjected to control systems have decreased so slowly. They also explain why pollution from agricultural chemicals such as nitrogen fertilizer, which arises from "nonpoint" sources, has continued to increase.

Thus, the decade or more of effort to improve the quality of the environment teaches us a fairly simple lesson: Pollution levels can be reduced enough at least to approach the goal of elimination only if the production or the use of the offending substances is halted. This precept directs our attention to the technology of production—the vast and varied machinery of industry, agriculture, and transportation. We can then see that all the really successful environmental improvements have been achieved by altering that technology. The very considerable reduction in mercury in Lake Erie nicely illustrates this fact. In 1970, there were very high concentrations of mercury in Lake Erie sediments; most of the lake bottom contained a thousand to two thousand parts per million, and more than that near the outlet of the Detroit River. In March of 1970, walleye contaminated with seven parts per million—fourteen times the acceptable level—were found in a tributary of the lake. The major source was soon discovered: chlorine-producing factories that used mercury to conduct the electric current that, passed through a brine solution, yielded the chlorine. Threatened by legal action, the plants adopted an alternate technology in which a semiporous diaphragm instead of mercury established the electrical circuit. By 1979, the concentration of mercury in most of the lake sediments was less than 300 parts per million, and fish levels were much improved. Mercury pollution in Lake Erie was sharply curtailed by the simple process of changing the technology of chlorine production.

PCB provides a similar example. This highly toxic synthetic substance was widely used in a number of processes, ranging from the manufacture of "carbonless" carbon paper to the manufacture of electrical transformers. When PCB production was banned, in the late 1970s, each of these manufacturing processes was simply altered to exclude it, and emissions into the environment fell precipitously. In the same way, the levels of DDT and related insecticides in the environment and in the human body have declined considerably because those substances were eliminated from agricultural practice in the early 1970s. In this case, the altered technology only replaced one hazard with another. Toxaphene, a hazardous insecticide, was widely used instead of DDT, especially for cotton crops; as a result, while their DDT content fell, the toxaphene content of fish increased more than twentyfold between 1970 and 1980.

In sum, there is a consistent explanation for the few instances of environmental success: They occur only when the relevant technologies of production are changed to eliminate the pollutant. If no such change is made, pollution continues unabated or, at best—if a control device is used—is only slightly reduced. For example, pollution from fertilizer nitrates, which cannot be controlled, has increased because nitrogen fertilizer, far from being banned, continues to play a major role in agricultural production. In the same way, adding a control device to the automobile engine without changing its basic technology has only slightly reduced carbon monoxide pollution. What has been done in regard to automotive technology accounts for both a striking environmental success, the 80 percent reduction in environmental lead, and a very troublesome failure, the continued emissions of nitrogen oxide, contributing to the attendant smog and acid rain. As we have seen, lead pollution was reduced by the simple expedient of cutting back its use in gasoline. This change was motivated partly by concern over the health hazard but perhaps more strongly by a technological accident: The platinum catalyst in the exhaust system designed to control carbon monoxide is poisoned by lead. But when the EPA was forced to mandate a shift from leaded to unleaded gasoline the lead had to be replaced, not merely eliminated, for it performed an essential function in current gasoline engines. It prevented the "knocking" that can seriously hamper the operation of modern, high-compression engines. Today, oil companies have replaced lead with petrochemical antiknock compounds (which they produce) or, to a much smaller extent, with ethanol produced from agricultural crops.

Removing lead from gasoline unequivocally eliminated its impact on the environment, but the nitrogen oxide problem is much more intractable. Unlike lead, which had been deliberately introduced into the gasoline engine, nitrogen oxides are an unwanted byproduct of its operation. Modern engines produce nitrogen oxides because they run at higher compression ratios, and therefore at higher temperatures, than older engines. At such elevated temperatures, when the oxygen and nitrogen in the air are drawn into the engine they react chemically, forming nitrogen oxides. After the Second World War, because American car manufacturers decided to build larger cars, they were forced to install more powerful engines. In turn, more power required a higher compression ratio, and, inevitably, nitrogen oxides were emitted. Thus, the auto industry's decision to build larger cars created the photochemical-smog problem and worsened the problem of acid rain.

Apart from a limited flirtation with small cars during recent energy shortages, the industry has been unwilling to build cars light enough to be driven by low-compression engines. An existing engine that can operate at high compression ratios without subjecting the air in the cylinder to excessive temperatures (the "charge-stratification" engine) could also help to solve the problem. But American manufacturers have thus far been reluctant to make the large-scale production changes needed to take advantage of this opportunity. Nitrogen oxides from automotive transportation could also be eliminated by the substitution of electric motors, which produce no environmental emissions at all, for the present ones. In diesel engines, diesel-fuel combustion produces carcinogenic chemicals, and these adhere to the tiny carbon particles emitted by the engine. Here, too, the auto industry has been unwilling to make the considerable changes in engine technology needed to reduce the hazardous emissions. Smog and diesel carcinogens continue to burden the air—and our bodies—because the productive technologies that generate them have not been changed to eliminate them.

In essence, the effort to deal with environmental pollution has been trivialized. A great deal of attention has been paid to designing—and enforcing the use of—control devices that can reduce hazardous emissions only moderately. Much less attention has been given to the more difficult but rewarding task of changing the basic technologies that produce the pollutants.

Production technology encompasses not only the design of a particular facility, such as a car, but the overall system in which it operates—in this case, transportation. Although remedying the vehicle's inherent faults is useful, much greater environmental improvement can be accomplished by addressing the system of transportation as a whole. Compared with railroads, cars are a highly inefficient means of long-distance and commuter travel, but they are essential in many places, even with effective mass transit, for urban travel. Cars designed exclusively for this purpose could be much lighter and driven by less powerful low-compression engines or, for that matter, by nonpolluting electric motors. Together with the expansion of urban and suburban mass-transit systems and interurban rail lines, this would virtually eliminate ozone and smog and greatly reduce environmental nitrogen oxides. Moreover, lower gasoline consumption would reduce carbon monoxide emissions without exhaust controls. In the same way, carcinogenic diesel exhaust could be sharply reduced by a shift from truckborne freight to railroad freight, which is four times as fuel efficient. In sum, the serious continuing impact of the transportation system on the environment results from the failure to alter not only the technical design of cars, trucks, and buses but also the system of transportation itself.

The effort to deal effectively with sulfur dioxide and dust pollution from power plants has been hampered by the same sort of superficial approach—reliance on control devices rather than changes in production technology. Since the early 1970s, more stringent requirements for control systems have apparently reduced the contribution of coal-burning power plants to sulfur dioxide and dust emissions. But changes in the overall structure of the power system have had the opposite effect. In recent years, coal has progressively displaced oil and natural gas as a power-plant fuel, and because these fuels produce far less sulfur dioxide and dust than coal does this shift has added to the sulfur dioxide problem. But no serious effort has been made to promote existing sources of power that can improve environmental quality. These include solar sources, such as photovoltaic cells, and decentralized power systems—notably cogenerators—that greatly enhance fuel efficiency, so that their environmental impact is much less than that of conventional power plants. Under the Reagan administration, government support for the development of such energy sources has been severely reduced.

In the petrochemical industry, which generates the toxic chemicals that so grievously pollute the environment, there is a unique relationship between environmental impact and the technology of production. In other sectors of production—the electric-power industry, say—the product is not itself a pollutant, and there are environmentally benign ways of producing it. In contrast, the manufacture of almost every petrochemical product is accompanied by the production of toxic wastes. Moreover, many of the goods manufactured, such as pesticides and plastics, are pollutants themselves. In effect, environmental pollution is an integral aspect of petrochemical-production technology.

The chemical disaster in Bhopal, India, in 1984, in which thousands of people died and 200,000 were injured, many of them for life, dramatizes the problems inherent in petrochemical-production technology. The conventional view is that the Bhopal disaster was due to some technical fault in the plant that produced the toxic chemical methyl isocyanate, which is used to manufacture insecticides. There is, of course, no perfectly accident-free industrial process, and this is particularly true of petrochemical plants, which not only are complex but turn out numerous dangerous products and byproducts. Indeed, there have been repeated accidents at the Union Carbide methyl-isocyanate plants both in Bhopal and in Institute, West Virginia—accidents that have exposed workers and in some cases the neighborhood to this dangerous substance. The conventional view accepts the inevitability of such risks and seeks to balance them against the benefits of producing methyl isocyanate. For example, a *Time* cover story on the Bhopal disaster stated:

> The citizens of Bhopal lived near the Union Carbide plant because they sought to live there. The plant provided jobs, the pesticide more

food. Bhopal was a modern parable of the risks and rewards originally engendered by the Industrial Revolution. . . . The world is Bhopal, a place where the occupational hazard is modern life. History teaches that there is no avoiding that hazard, and no point in trying; one only trusts that the gods in the machines will give a good deal more than they take away.

This is a widely held position. The chemical plant improved the life of the Indian people; such disasters seem to be a price that must be paid for such benefits of modern production technology. But the facts contradict this position. Consider how well the Bhopal plant "provided jobs." Bhopal is a city of 670,000 people, and perhaps a third of them were living in shanties next to the plant. But the plant provided jobs for only a few hundred. The community that grew up around the plant had been attracted not by jobs but by secondary amenities like the water supply and the roads, and the vast majority of the people lived in extreme poverty.

And how well do the pesticides that are produced from methyl isocyanate—carbamate insecticides such as Sevin and Temik—provide more food? Such pesticides can, of course, protect crops from insect attack and improve production. However, in developing countries most pesticides are used not on local food crops but on crops grown for export—cotton, coffee, bananas. Such "monoculture" agriculture is particularly susceptible to insect attack, so there is a heavy reliance on pesticides. For example, a 1979 study reported that El Salvador, a very small country, used about one fifth of the world's parathion output to treat its coffee crop. Moreover, Third World countries use many hazardous pesticides that have been banned in the developed countries. Although relatively little pesticide is used directly on food crops, they are nevertheless contaminated by pesticides sprayed on nearby export crops. As a result, food grown in Third World countries is often heavily contaminated. Yet crops can be produced with much less dependence on pesticides. "Integrated pest management" is a way of deciding at precisely what time and in precisely what amounts an insecticide is needed. It can considerably reduce the amounts used, and therefore the danger of contamination. Because of opposition from the chemical companies, however, this process is still little used in the United States or in other developed countries and is used hardly at all in developing countries.

The real goal at Bhopal, therefore, should be not merely to make the plant safer but to end the need for operating it by transforming the technology of agricultural production: eliminating—or, at least, greatly reducing—its dependence on pesticides. Rather than trust in the gods in the machines, we need to decide for ourselves what kinds of machines best serve both the environment and society.

Unlike the steel, auto, and electric-power industries, the petrochemical industry—on its present scale, at least—is not essential. Nearly all its products are substitutes for perfectly serviceable preexisting ones; plastics for paper, wood, and metals; detergents for soap; nitrogen fertilizer for soil, organic matter, and nitrogen-fixing crops (the natural sources of nitrogen); pesticides for the insects' natural predators. Apart from relatively few items that cannot be produced in any other way—such as pharmaceutical drugs, videotape, and the plastic artificial heart—petrochemical products could be replaced by less hazardous ones. Thus, the petrochemical industry is unique: Not only its wastes but its very products degrade the environment, its hazards are largely immune to either prevention or control, and most of its products are replaceable. The petrochemical industry is inherently inimical to environmental quality. The only effective way to limit its dangerous impact on the environment is to limit the industry itself.

To ardent environmentalists, such talk about the technology of production may seem an irrelevant, or even a hostile, digression. Like any embattled minority, environmentalists are prone to emphasize the singular importance of their special interest—the preservation of the natural environment. This position, of course, has a good deal of merit; after all, every aspect of human society depends on the ecosystem—for the natural resources that support production, for the air we breathe, for the green plants that feed us and other animals. Moreover, the environment is governed by stubborn, largely unalterable natural forces, whereas the system of production is subject to human choice. Logically, therefore, the decisions that determine the choice of production technology ought to be governed by the constraints inherent in nature. But in fact the actual direction of governance is reversed. The design of the manmade system of production is not governed by the nicely balanced attributes of the natural system; instead, as we have seen, the state of the environment is determined by the production technologies that have been chosen to support industry, agriculture, and transportation.

So the environmentalist who wishes to grapple with this illogical arrangement needs to turn from the fairly rigid but harmonious pattern of nature to the more flexible but chaotic realm of human decisions. And this realm necessarily includes not only the choice of production technologies but also the closely related economic decisions. Logic tells us that the economic system—the processes that mediate the flow of wealth—is contingent on the system of production, which is, after all, the source of the goods and services that generate economic wealth. In a logical arrangement, therefore, just as environmental constraints ought to govern the design of the production system, that design should, in turn, specify appropriate features of the economic system. For example, if the importance of eliminating smog ought to specify the design of automobile engines, then the need to build such engines ought to govern the automobile industry's investment decisions—in part, at least. But here, too, the logical direction of governance is reversed: The decisions that determine the design of the system of production—what is produced and by what technological means—are determined largely by economic considerations that are quite independent of the product's environmental impact, the chief consideration being the profitability of the enterprise.

In sum, the decisions that govern environmental quality originate in the economic realm and are translated into the design of productive technology, which—as now constituted—in turn visits upon the environment the evils of pollution. This relationship becomes clear when we consider the origins of the major environmental problems. Smog, for example, is the end result of the economic motivation that led the auto industry to decide to manufacture large, powerful cars, for, as John DeLorean, a former General Motors executive, has said, "when we should have been planning switches to smaller, more fuel-efficient, lighter cars in the late 1960s in reponse to a growing demand in the marketplace, G.M. management refused because 'we make more money on big cars.' "

Recent changes in the technology of electric-power production that have seriously affected environmental quality have also been economically motivated. When the utilities were persuaded that nuclear power would be "too cheap to meter," they hastened to build nuclear power plants, thereby creating the monumental hazards of nuclear accidents and the still unresolved problem of radioactive waste. Then, when this hope turned into an illusion, there was a rapid switch to coal-fired plants, because they produce cheaper electricity than oil-powered or nuclear-powered plants—but they also produce much more sulfur dioxide. In the same way, the steel industry's response to the demand for plants that would be less polluting—closing rather than improving them—is motivated by the higher profits obtainable from alternative investments in oil and chemicals. Similarly, the postwar history of American agricultural chemicals was motivated by the increased economic returns on investment that they engendered. Finally, the production of petrochemicals is more profitable than the manufacture of the products that they replace; for example, profits on detergent production are significantly higher than profits on soap production.

It is economic motivation, then, that has impelled the sweeping anti-ecological changes in the technology of production that have occurred since the Second World War. These changes have turned the nation's factories, farms, vehicles, and shops into seedbeds of pollution: nitrates from fertilizer; phosphates from detergents; toxic residues from pesticides; smog and carcinogenic exhaust from vehicles; the growing list of toxic chemicals and the mounds of undegradable plastic containers, wrappings, and gewgaws from the petrochemical industry.

The decisions to make these technological changes have been short-sighted not only because of their impact on the environment but also because they often generate serious economic problems. For example, the intensive use of agricultural chemicals has harmed not only the environment but, in recent years, the agricultural economy as well. As increasing amounts of fertilizer are applied to the soil, crop growth reaches a limit, and thereafter the yield produced per unit of fertilizer falls and economic productivity is reduced. At the same time, excess, unassimilated fertilizer leaches through the soil into surface water and intensifies pollution. Similarly, the continued use of pesticides breeds resistance in the pests, so that more and more of the chemicals must be used to achieve the same effect. For these reasons, the economic productivity of agricultural chemicals—that is, the crop output per unit of chemical input—has decreased by 50 percent since 1960, and it is still falling. Thus, both increased stress on the environment and falling economic efficiency are inherent in the present system of agricultural production. In the past few years, American farming has been undergoing a deepening economic crisis, marked by numerous bankruptcies, especially of family farms. Each year, as agricultural chemicals become more expensive and less efficient in improving crop production, farmers must incur a higher debt to buy the chemicals and thereby become more vulnerable to bankruptcy. The heavy dependence of agricultural technology on chemicals has hurt not only the quality of the environment but the farmers' livelihood as well.

Economics also explains the reluctance of the petrochemical industry to deal effectively with its toxic waste. The only practical, though far from satisfactory, way to keep toxic chemicals out of the environment is to destroy them—in specially designed incinerators, for example. This process, however, is so expensive—as much as $2,000 per metric ton—that if it were to be applied to the total annual tonnage of toxic chemicals (an estimated 2,650,000 tons) it might cost about $5.5 billion, an amount larger than the annual profit of the thirty largest American chemical companies, which include most of the nation's petrochemical industry. Proper treatment of its waste would thus force the industry to raise prices and face serious competition from the natural products that it has replaced.

In the end, the decisions that have so gravely affected the environment have also had serious economic effects, because they govern a factor that is closely linked to both the environment and the economy—the system of production. In a free-market economy, capital tends to be invested in those production enterprises that promise to yield the greatest return in the shortest time. Because the social interest in environmental quality or long-term economic efficiency is not represented in such investment decisions, neither of these desirable results is likely to be achieved by them.

There seems to be a paradox embedded in the conclusion that most, if not all, environmental pollution is the outcome of decisions guided by free-market forces. It is often argued, after all, that the market operates—automatically, in the best of circumstances—to seek out those decisions that make the most efficient use of the available resources. And one of the well-justified claims for ecological processes is that *they* are characteristically efficient: There are, after all, no "wastes" in the ecosystem. If the present system of production is counterecological, it must be wasteful; if this is so, new systems of production, more closely guided by ecological principles, ought to be more efficient than those devised by the present market-guided system. Why then haven't they attracted investors and entered into successful competition with the present system?

The overall economic efficiency of the new technologies that have transformed production since the Second World War—and have polluted the country—is unmistakable. Considerably more profitable than the technologies they have replaced, the new ones have generated a more than fourfold increase in total production (as measured by the gross national product) since 1950. However, the improvement in efficiency responsible for this achievement has been unbalanced. Although the efficiency with which labor is used—that is, the value of the product yielded per hour of a person's labor—has increased a great deal, the corresponding productivity of capital and natural resources such as energy has declined. For example, the postwar substitution of plastics for leather has displaced an industry characterized by low labor productivity and high capital productivity with an industry characterized by high labor productivity and low capital productivity. To generate the same value of product, the plastics industry uses about a fourth as much labor but ten times as much capital as the leather industry. The disparity between the two industries' energy productivity—the amount of energy used to produce a given value of product—is even greater: The plastics industry uses about thirty times as much energy as the leather industry.

However, if one takes a more fundamental approach to the problem of environmental quality by recognizing that it is inherently linked to the technology of production, one can find ways of improving both the economy and the environment. The electric-power industry provides an informative example. The power industry is a notorious source of pollution: responsible for a great deal of environmental dust, sulfur dioxide, and nitrogen oxide; contributing considerably to acid rain; and generating the threat of nuclear disasters. In the United States, the industry is also in a precarious economic condition: The utilities' heavy investment in huge power plants—especially nuclear plants—has driven the cost of producing electricity sharply upward and has threatened some of the companies with bankruptcy. (A *Business Week* cover story has asked the question "Are

Utilities Obsolete?") A major reason for these economic difficulties is that the present technology of electric-power production is highly centralized. Typically, each new plant is very large and requires a huge capital investment—billions of dollars. The plant is so large because it represents an investment in the capacity needed well into the future. Consequently, when a new plant begins to operate, the system inevitably has an excess capacity, and for a time part of the capital investment yields little or no return. Moreover, the transmission system that distributes electricity from central stations is costly and consumes a significant fraction of the power. Finally, centralized power plants are inherently inefficient, because, for inescapable thermodynamic reasons, two thirds of the energy available from the fuel is dispersed into the environment. This means that the industry wastes two thirds of the fuel that it uses and causes three times as much pollution per unit of useful energy produced as it would if the wasted energy could be recaptured for use.

The heat discarded by a central power station could readily be recaptured and used—to heat homes, for example. But large centralized plants cannot be used for this process—cogeneration—because heat can be transmitted effectively only over short distances. No one wants to live such a short distance from a power plant, especially if it is nuclear. (Some people are in this unhappy position. For example, a nuclear power plant in Gorki, in the Soviet Union, provides heat for that city; its presence must have been causing a good deal of concern since the Chernobyl disaster.) By redesigning the technology of electrical production in keeping with sensible ecological principles, both environmental and economic improvement can be achieved. What is needed is to decentralize the production of electricity and install cogenerator plants just large enough to meet the local demand for heat and electricity. With the cogenerator operating at a much higher level of both economic and thermodynamic efficiency than a conventional power plant, fuel consumption is reduced, and the environmental impact and the cost of energy are decreased.

Similar environmental and economic gains could be achieved by reorganizing agricultural production along ecological lines. For example, it has been found that the net economic returns of large-scale Midwestern organic farms, which use no fertilizer or pesticides, are equal to the returns of otherwise similar conventional farms, which use a great deal of these agricultural chemicals. This represents an economic gain for the organic farmer, because with lower production costs they are less dependent on bank loans and therefore less vulnerable to bankruptcy. Similarly, if the crop system were properly redesigned to produce both food and ethanol (a solar fuel, and thus renewable), the farmers' income could be diversified and national energy production enhanced.

These gains can be achieved only if the social need for both environmental quality and economic growth is allowed to govern the choice of production technology. Typically, measures designed to meet such a need require a relatively large initial investment. This is a serious hurdle, which people who would benefit most—the poor—have difficulty in overcoming. Another hurdle is the relatively slow rate of return on some of the necessary investments. If such an investment—for example, the relatively slow restoration of impoverished soil through a transition of organic farming—must compete with an existing investment that yields quick profits, it is certain to be ignored. However valuable to society, the investment is not likely to be made if it must compete in the free market, which favors large-scale investments over decentralized ones and short-term over long-term returns.

Perhaps the most profound question raised by environmental issues is to what extent the choice of production technologies should be determined by private economic considerations and to what extent by social concerns like environmental quality. These values are in sharp conflict. There appears to be a broad consensus that it is in the national interest to restore the quality of the environment, and the resultant legislation confirms the general impression that achieving this goal is a social, governmental responsibility. And, as we have seen, significant environmental improvement requires the proper choice of technologies and systems of production, so that this choice, too, becomes a social responsibility. Yet in our free-market economy the right to make such a choice is in private, not public, hands, and this is a right that very few Americans would challenge.

The recent entrepreneurial history of the United States Steel Corporation (now renamed USX) illustrates what can happen in these circumstances. In the 1970s, the company developed a plan to build a large, modern, low-pollution steel mill on Conneaut Lake, in Pennsylvania—a socially useful way to reduce both steel imports and pollution. Then, in 1982, with the price of steel driven down by cheap imports and the price of oil rising, U.S. Steel decided to buy the Marathon Oil Company, for about $6 billion. The Conneaut Lake plant was never built. By 1985, 54 percent of the company's business was in oil and gas and only 35 percent in steel. In July of 1986, the company dropped the word "Steel" from its name in favor of the ambiguous "X." The economist Robert Reno has commented:

> Now, it is arguable whether the best interest of the nation would have been better served if U.S. Steel had reinvested in modernized steel plants rather than purchasing Marathon Oil. Anyway, there's no law that says corporations must choose national interests over those of their stockholders.

In fact, the company's legal obligation is not to the nation but to its stockholders, and that obligation is to produce not steel but profits.

This is, after all, the meaning of "free enterprise": The owner of capital is at liberty to invest it in any enterprise—whether it produces steel, chemicals, or oil—that offers the investor the most promising rate of return, the greatest market share, or some other private advantage.

The institution that largely determines the course of production technology in the United States—the corporation—is itself a peculiar amalgam of social and private features. In one sense, a basic purpose of the corporate form is social: to accumulate from a relatively large number of people an amount of capital far larger than that available from any one person. Moreover, almost all the earliest American corporations were chartered for a specific social purpose—constructing a canal, a bridge, a turnpike, or establishing a bank—that required large, collective capitalization. Today, corporations may be chartered for any legal purpose, and thus have the right to determine privately how their socially collected capital will be invested. However, corporations are certainly still social in their effects—not only on the environment but on employment, wages, and working conditions as well. Yet with respect to their control, corporations are private, for they are governed by a small, self-perpetuating group of officers, who have no general accountability to society except for adherence to relevant government regulations, or—apart from profitability—to the stockholders who own the corporation.

The conflict between the modern corporation's social role and its private control led Adolf A. Berle and Gardiner Means to declare, in their classic work *The Modern Corporation and Private Property*:

> The economic power in the hands of the few persons who control a giant corporation is a tremendous force which can harm or benefit a multitude of individuals, affect whole districts, shift the currents of trade, bring ruin to one community and prosperity to another. The organizations which they control have passed far beyond the realm of private enterprise—they have become more nearly social institutions.

From their detailed analysis of this situation, Berle and Means concluded that the separation of corporate ownership from control places society "in a position to demand that the modern corporation serve not alone the owners . . . but all society" and that "it remains only for the claims of the community to put forward with clarity and force."

The date of Berle and Means' work—1932—is significant, for it suggests that the shattering impact of the Great Depression on confidence in the free-market system had impelled them to seek alternatives. Since then, potential critics have been, if not silenced, at least inhibited by the remarkable vitality of the American economy after the Second World War: by the dazzling new products and by the growth of national and private wealth. Now that the attendant social ills—persistent unemployment, insufficient affordable homes, pollution, and the threat of nuclear annihilation—have become more evident, the silence has been broken. For example, the recent Catholic bishops' pastoral letter on the economy, citing Thomas Aquinas, asserts, "No one can ever own capital resources absolutely or control their use without regard for others and society as a whole. . . . Short-term profits reaped at the cost of depletion of natural resources or the pollution of the environment violates this trust." The bishops' letter reflects the 1981 encyclical "On Human Work," in which Pope John Paul II asserted, on moral grounds, that ownership does not justify exclusive control of production facilities: ". . . in consideration of human labor and of common access to the goods meant for man, one cannot exclude the socialization, in suitable conditions, of certain means of production." One can, of course, embellish these observations and arguments with a doctrinal term—socialism—and embrace or dismiss them by reacting to that term. While this may satisfy one's ideological convictions, neither approval nor disapproval can alter the reality, which, as we have seen, is that substantial environmental improvement has occurred only when the choice of productive technology was open to social governance.

The New Deal's response to Berle and Means' statement concerning the "claims of the community" on the corporations was regulatory legislation, then applying especially to banking and securities. This served as a model for subsequent regulatory measures, which reached a peak in the environmental legislation of the 1970s. These social interventions into corporate activity have been proscriptive rather than constructive: They tell the corporations what not to do rather than what should be done. Yet certain provisions of the new environmental legislation do in fact at least imply that environmental improvement may be achieved by the socially mandated choice of production technology. Thus, the basic instrument created by environmental legislation, the environmental impact statement, is required to include a presentation of "alternatives to the proposed action" when that action would have an unavoidable adverse environmental effect. This provision implies that in, say, a public hearing on a trash-burning incinerator that would unavoidably emit dioxin into the environment, a citizen could call for a decision mandating the choice of a technology that would separate and recycle trash rather than burn it. The banning of the use of DDT in agriculture or of PCB in manufacturing is a concrete, if relatively narrow, example of this approach. Opportunities, then, do exist within the framework of environmental legislation that, though they have been little used thus far, would permit at least the public weighing of alternative choices of production technologies.

Since conventional environmentalism is preoccupied with controlling emissions rather than with changing the production processes that generate them, economic considerations are usually reduced to what is called a cost-benefit analysis: The cost of the necessary controls is compared with the value of benefits (to human health, for example) generated by a reduction in the level of pollution. Such an analysis attempts to give a social aim—the protection of the environment and of public health—dollar dimensions. Transformed into a pseudoeconomic value, the social value can then be compared, it is supposed, with other economic factors, such as the cost of controls, and some "cost-effective" balance struck.

An illuminating example of the moral booby traps built into the cost-benefit philosophy is provided by the EPA's recent justification for further reducing the lead allowed in gasoline. The EPA carried out a cost-benefit analysis by totting up the cost of more expensive methods to boost octane and refinery changes needed to replace lead in gasoline (about $600 million in 1986) and comparing it with the monetary value of less engine damage from the lead additives, better fuel economy, reduced emissions of other conventional pollutants, and savings in the cost of medical care and compensatory education for children suffering from the physical and cognitive effects of lead poisoning (a total of about $2 billion in 1986). The EPA analysis concluded that the dollar value of the benefits of reducing lead in gasoline was more than three times the cost—clearly a bargain to be snapped up. In those terms, the decision seems to have no more moral content than choosing one brand of breakfast cereal over another. Suppose, however, that a new medical advance were to reduce sharply the cost of treating lead poisoning, and so bring the dollar value of the benefits below the cost of controlling lead. What then? Would this new cost-benefit balance justify continued lead pollution? The apparently "objective" cost-benefit approach is thus quickly engulfed in deep moral issues. Should society attempt to mitigate *all* human suffering, regardless of cost? On the other hand, as social resources are limited, is it not reasonable to alleviate suffering in keeping with some measure of effectiveness? If a monetary measure is objectionable, what better measure can be used? It is, of course, unreasonable to ask the EPA to answer these questions, which have frustrated numerous efforts over many years to answer them. But it is not unreasonable to insist that these issues be acknowledged rather than reduced to a trivial, easily dismissed monetary "solution."

Risk-benefit computations involving toxic pollutants have a similar outcome. Because the pollutants' ultimate effect can often be assessed by the number of lives lost (from cancer caused by an environmental carcinogen, for instance), the risk-benefit analysis requires that a value be placed on a human life. Some economists have proposed that the value should be based on a person's lifelong earning power. It then turns out that a woman's life is worth much less than a man's, and that a black's life is worth much less than a white's. In effect, the environmental harm is regarded as smaller if the people it kills are poor — a standard that could be used to justify situating heavily polluting operations in poor neighborhoods. And, in fact, this is an all-too-common practice. Thus, thinly veiled by seemingly straightforward numerical computation, there is a profound, unresolved moral question: Should poor people by subjected to a more severe environmental burden than others, simply because they lack the resources to evade it? Once again, what is at issue is not to resolve this question but to allow that it be asked. These examples warn us that environmental considerations are likely to raise troubling social issues — issues that in recent practice have been swiftly reduced to "solutions" by forcing them into the Procustean bed of the free-market economy.

Recognition that significant environmental improvement depends on social rather than private governance of production decisions helps us understand why the considerable effort to improve the environment has had so little effect. This effort, however forceful, meets a politically immovable object: the conviction, powerfully embedded in American society, that the decisions that determine what is produced and by what technological means ought to remain in private, corporate hands. In the United States, there appears to be a powerful taboo against even public discussion, let alone criticism, of this principle. Confronted by the taboo, the environmental movement has splintered into a series of separate, more palatable campaigns.

Another consequence of this unresolved confrontation is that environmental organizations have been drawn toward explanations of the environmental crisis that avoid its origin in economic forces and appeal instead to the principles of ecology. Thus, on the assumption that the global ecosphere is a closed system of limited capacity, some environmentalists claim that the key to environmental quality is population control. Similarly, on the assumption that ecological cycles are closed and self-sustained, other environmentalists maintain that people, too, should abide by this rule, and live in communities supported by their own local or regional resources. And others derive from these same considerations the conclusion that ecological improvement requires the imposition of controls on economic growth.

The "limits to growth" approach is based on a serious misconception about the global ecosystem. It reflects the notion that the earth is like a spaceship—a closed system, isolated from all outside sources of support and necessarily sustained only by its own limited resources. The earth, however, is not a closed, isolated system, for its behavior is totally dependent on the huge influx of energy from an outside source—the sun. (In fact, the same is true of the spaceship, for it relies on sunlight to generate the electricity that runs its machinery.) On the earth, solar energy creates the weather: the seasonal and daily changes in temperature; the moisture that the sun lifts from the oceans; the storms that carry rain and snow to the earth and replenish the lakes and rivers that feed the oceans. The weather, in turn, molds the physical features of the earth's surface, creating the ecological niches that living things occupy. Solar energy, captured by photosynthesis, sustains every form of life and drives the ecological cycles in which they participate. If an ecological cycle is viewed only as a static array of animals, plants, and microorganisms linked through the physical environment into a circular system, it appears to be closed, like a ring. But this image is misleading, for without the energy that the ecosystem receives externally, from the sun, the plants and animals would die; the circular system would disintegrate.

This misconception also leads to the notion that the earth, as a supposedly isolated system, is subject to the increasing chaos arising from the unavoidable increase in entropy. Entropy is a measure of disorder in a physical system, which is bound to increase whenever energy is used to generate the system's only useful product, work. But this increase in disorder happens only in a system that is isolated from an outside source of energy. The fact is that in the earth's ecosystem entropy-increasing processes—for example, the combustion of fuel, the death and decay of living organisms—are reversed by the external supply of solar energy, which regenerates burnable organic matter through photosynthesis. The organic matter, in turn, provides the energy that drives the entropy-decreasing process of plant and animal growth. On the earth, solar energy counteracts entropy.

In the abstract sense, there is a limit to economic growth, yet this limit is determined not by the present availability of resources but by a distant limit to the availability of solar energy. It is true, of course, that the ecosystem occupying the earth's thin skin and the mineral deposits underlying that skin are essential to economic production and that they are finite. It is also true that economic growth is limited by the finite amounts of those essential resources. However, because mineral elements are indestructible they can be recycled and reused indefinitely as long as the energy necessary to collect and refine them is available. And precisely this is done when the resource is sufficiently valuable: Despite extensive dispersion, well over half of all the gold ever mined is still in hand today. Hence the ultimate limit on economic growth is imposed by the rate at which renewable solar energy can be captured and used. If we ignore the exceedingly slow extinction of the sun, this limit is governed only by the finite surface of the earth, which determines how much the energy radiated by the sun is actually intercepted and is therefore capable of being used. How close are we to this limit? It has been estimated that the solar energy that falls annually on the earth's land surface is more than a thousand times the amount of energy (from fuels and hydroelectric and nuclear power) now being used annually to support the global economy. Of course, because of the relative inaccessibility of some parts of the earth not all the solar energy that reaches the planet could be used. If, let us say, only 10 percent of the solar energy falling on the land could be captured, it would still be possible to expand our present rate of energy use perhaps a hundredfold before encountering the theoretical limit to growth. Even if this figure should turn out to be somewhat optimistic, it seems clear that at present we are nowhere near the limit that the availability of solar energy will eventually impose on production and economic growth. That distant limit is simply irrelevant to current policy decisions. The issue we face, then, is not how to facilitate environmental quality by limiting economic development but how to create a system of production that can grow and develop in harmony with the environment.

There have been lively debates over whether environmental degradation can be reversed by controlling the growth of the world population. A decade ago, many if not most environmentalists were convinced that population pressure and "affluence," rather than inherent faults in the technology of production, were the chief reasons for environmental degradation. Since then, the hazards dramatized by Chernobyl, Three Mile Island, Seveso, Bhopal, Love Canal, and Agent Orange have convinced many people that what needs to be controlled is not the birth rate but the production technologies that have engendered these calamities. In the last few years, however, the notion of population control has been given a new impetus, not so much from popular support as from the declarations of national environmental organizations and some of their leaders. Several examples are worth noting. Richard D. Lamm, the ex-governor of Colorado, who is a leading environmentalist, has frequently pleaded for population control, demanding an end to the flow of illegal immigrants into the United States from what he calls "never-to-be-developed countries," and urging on the terminally ill a "duty to die." In support of these ideas, he cites Aristotle: "From time to time it is necessary that pestilence, famine, and war prune the luxuriant growth of the human race." Garrett Hardin, a frequently cited environmentalist who is also a major proponent of population control, once gave this very general, ancient proposition an apparently ironic, distinctly modern — if astonishing — form:

> How can we help a foreign country to escape overpopulation? Clearly, the worst thing we can do is send food. . . . Atomic bombs would be kinder. For a few moments the misery would be acute, but it would soon come to an end for most of the people, leaving a very few survivors to suffer thereafter.

For a more reasoned support of population control, we can turn to Russell W. Peterson, the former president of the Audubon Society, which is a major environmental organization:

> Almost every environmental problem, almost every social and political problem as well, either stems from or is exacerbated by the growth of human population. . . . As any wildlife biologist knows, once a species reproduces itself beyond the carrying capacity of its habitat, natural checks and balances come into play. . . . The human species

is governed by this same natural law. And there are signs in many parts of the world today—Ethiopia is only one of many places, a tip of the iceberg—that we *Homo sapiens* are beginning to exceed the carrying capacity of the planet.

Examples can be readily cited to show that Peterson's conclusion is wrong, and that, quite apart from moral considerations, Hardin's Swiftian proposal to use nuclear weapons as a contraceptive device is therefore misguided. It is useful to remember that people in other countries did not go hungry because they sent food to Ethiopia to relieve the famine— that what was sent to Ethiopia was *surplus* food. In fact, the world produces more than enough food to feed the total world population. Total world production of food, equally distributed to the global population, would today provide everyone with more than enough for the physiologically required diet. According to a recent estimate by the United Nations Food and Agriculture Organization, the world now produces enough grain alone to provide every person on earth with 3,600 calories a day—more than one and a half times the calories required in a normal diet. Enough grain is produced to give everyone on earth two daily loaves of bread.

Famine is caused not by global food shortage but by the grossly uneven distribution of the global food supply. This is not an ecological phenomenon but a political and economic one. Neither England nor Haiti produces enough food for its own population, but hunger is much more prevalent in Haiti than it is in England, because Haiti cannot afford to import enough food to make up this deficit, whereas England can. A more localized example of the economic origin of hunger comes from a 1967 study of the state of Madras, in India. The study found that about a third of the population consumed an amount of food well below the physiological requirements for protein and calories: Those people were slowly starving. But the total food output of the state was sufficient to provide the entire population with 99 percent of the required diet. The problem was not an ecological imbalance between the size of the population and the size of the food supply. The problem was poverty: The well-fed people earned about 50 percent more than the starving ones. In general, there is considerable evidence that in developing countries poverty encourages a high birth rate, for parents hope that enough children will survive the scourge of disease to support the family. The reason for malnutrition, starvation, and famine, then, is poverty, not overpopulation. Excess population is a symptom of poverty, not the other way around. Hunger and overpopulation are not ecological manifestations; they are signs of economic and political problems that can be solved, humanely, by economic and political means.

Many people have been attracted to the environmental movement by their dismay over the present state of human society. Some of them are interested not only in correcting the environmental abuses that erode the quality of life but also in establishing a new, happier way of living. They want to create better places to live and better ways to work. They regard industrial society as the root not only of environmental problems but of all social, political, and economic ones as well. They look toward ecology as a guide to the creation of a new culture, a new style of living. Kirkpatrick Sale, a prominent exponent of this view, wrote in *The Nation* that this style of living would be "rooted in the natural world, in harmony with natural systems and rhythms, constrained by natural limits and capacities and developed according to the natural configurations of the earth and its inherent life forms." Seeking to establish an ecologically harmonious way of life—for themselves, at least—some people return to the land, growing their own food, building their own homes and furniture, recycling their waste to the land. Others have a wider vision, seeking "self-reliance, not so much at the individual as at the regional level." This approach, "bioregionalism," envisions a society based on ecologically defined regions rather than on areas specified by political boundaries. It embraces the belief that such an ecologically founded society will have many laudable features: It will be cooperative rather than competitive, spread out rather than centralized, interdepenent rather than polarized, evolving rather than growing, peaceful rather than violent.

Certainly it makes ecological and, ultimately, economic sense to organize society in ways that harmonize with nature. Problems arise, however, when one attempts to translate this general proposition into practice. According to Sale, bioregionalism is achievable, because its appeal overrides conventional political distinctions, and thereby avoids polarization and opposition:

> The bioregional idea has the potential to join what are traditionally

thought of as the right and the left in America because it is built on and appeals to values that, at bottom, are shared by those on both sides.

In effect, Sale hopes bioregionalism can eventually remake society and somehow overcome opposition from the powerful economic forces that now largely govern it. This would be a remarkable political conjuring trick, for there is no way to reorganize society along ecologically sound lines without directly challenging the powerful, politically conservative forces—more plainly speaking, the corporations—that now control the system of production.

The illusion that environmental improvement is a politically neutral issue is not new. When environmental quality first emerged as a public concern, in 1970, a major politician declared, "Ecology has become the political substitute for the word 'motherhood.' " The theory that conservatives and liberals will happily join forces to develop an ecologically sound society has been tested by an eminent practitioner of political realism, Richard M. Nixon. He discovered that even his administration's modest environmental efforts were a serious threat to corporate power. In 1970, on signing the National Environmental Policy Act, President Nixon adopted the environmental issue as his own, promising a vigorous campaign to clean up pollution. He said, "The nineteen-seventies absolutely must be the years when America pays its debt to the past by reclaiming the purity of its air, its waters, and our living environment. It is literally now or never." In his 1971 State of the Union message, Nixon, still sounding like a crusading ecologist, declared that environmental quality would be the "third great goal" of what he called the New American Revolution. But less than a year later Nixon seemed to believe that environmentalism threatened the basic precepts of corporate power. In September of 1971, he indicated that he had made a complete reappraisal of his environmental policies, and told an audience of auto-industry executives that he would not permit concern for the environment "to be used sometimes falsely and sometimes in a demagogic way basically to destroy the system."

In referring to the environmental impact of production technologies, environmentalists often speak of the "hard path" and the "soft path." The hard path—the huge, centralized technologies created without concern for their environmental impact—leads to nuclear power, chemical agriculture, and high-powered cars. The soft path leads to technologies more appropriate in scale and design to their human purpose and ecological setting—cogenerators, solar energy, organic farming and small cars, mass transit and bicycles. This distinction is useful, but it is incomplete; in reality, the soft path of ecologically appropriate technology is a hard political road, for those traveling it confront the real source of environmental degradation—the technological choice—and debate who should determine that choice, and for what purpose. And a soft political road, which avoids this debate, inevitably leads to the environmental hazards presented by the giant technologies.

In the United States, the major environmental organizations are aware that in order to influence the resolution of environmental issues they need to find a way of participating in the political process. But they have generally chosen to travel the soft political road, which evades the issue of who should govern the choice of productive technology. This is evident in the recently published *An Environmental Agenda for the Future*, by a group of ten self-described "leaders of America's foremost environmental organizations." The agenda recognizes the need to confront the root causes of environmental degradation, stating that "because the laws and regulations often have not dealt with root causes, they have been inadequate to cope with added problems that have arisen, partly from new technologies." Given this laudable intention, and what we now know about the root cause of environmental pollution, one would expect the environmental leaders to guide us toward the future by proposing national policies such as these: that the present reliance on nonrenewable energy sources, and especially nuclear power, be replaced by renewable solar-energy systems; that the present chemically based system of agriculture be replaced by a system based on organic farming; that the inherently dangerous petrochemical industry be rolled back, its products and processes replaced by ones more compatible with the environment; that the nation's present heavy dependence on cars and trucks be replaced by a vast expansion of railroads and mass-transit facilities. However, the agenda offers a policy statement only with respect to population control: "The Administration should establish formal population policies, including goals for the stabilization of population at a level that will permit sustainable management of resources and a reasonably high quality of life for all people." Indeed, the agenda takes a gingerly approach to any measure that calls for socially mandated changes in the technology of production. For example, although favoring a shift toward renewable energy, the agenda recommends only that federal and state governments "develop successful strategies," rather than recommending a formal policy, as in the case of population. Similarly, according to the agenda, organic farming should be "researched and encouraged," but no political action is proposed actually to put this advantageous technology into practice.

The agenda appears to reflect the environmental leaders' effort to adjust the goals of environmentalism to the reality of Washington politics. As they see it, the reality is this: To obtain legislation that would improve environmental policy—by urging more stringent pollution standards, for example—a majority of Congress must be won over. But legislators are subjected to the influence of well-financed industrial lobbyists, who fiercely resist such restrictions on their companies' operations. In the past, these battles have often been lost by the environmentalists or, at best, have produced a legislative compromise. The alternative to lobbying is litigation—a lengthy, costly, and often inconclusive process. Thus far, clearly, neither effort has done much to improve the state of the environment.

In recent years, with an administration that strives for minimal government programs in all but military affairs, environmental lobbying has seemed particularly fruitless. The environmental organizations, which have found themselves facing mounting costs and diminishing returns, have accordingly shown a tendency to seek less frustrating—and less costly—routes to environmental improvement. The main outcome has been the formation of small committees, composed of leaders of environmental organizations and corporate executives, that seek to work out compromise positions.

A good example is the Acid Rain Roundtable, a twelve-member group representing environmental organizations and power companies. In 1985, the Roundtable reported the results of a study that it had commissioned to devise "cost-effective programs" for the reduction of emissions of sulfur dioxide and nitrogen oxides from power plants. The study concludes that a reduction of 26 percent in nitrogen-oxide emissions could be achieved at an annual cost of about $2.6 billion dollars. The study opposes the strategy of reducing sulfur dioxide emissions by installing scrubbers, which are regarded as too costly. According to the study, a more cost-effective solution is to cut back the burning of high-sulfur coal—a step that would reduce coal mining in the Midwest and northern Appalachia. Concerning the resultant unemployment, one Roundtable participant notes that the study was directed toward ways "to address miner unemployment by means other than mandating the installation of scrubbers." The report proposes to aid the miners who would lose their lifelong jobs by helping them obtain unemployment and health insurance for one year after they are thrown out of work. Regardless of the merits of the compromise recommended by the study, its basic structure is clear: It relieves the power companies of the cost of installing scrubbers, or of the competition threatened by local cogenerators—both facilities capable of reducing pollutant emissions—and shift the economic consequences of achieiving a modest reduction in emissions onto the miners. According to one Roundtable member, the study was designed to "break the legislative deadlock." Not surprisingly, the United Mine Workers is not one of the member organizations of the Roundtable, although, as important parties to the issue, they would receive respectful attention if it were to be resolved by conventional democratic means, in the legislature.

One result of the national organizations' tendency to travel the soft political road has been the appearance of a new grassroots environmental movement. This movement has arisen out of the direct impact of environmental pollution on community life. It is concerned chiefly with toxic chemicals, toxic-waste dumps, and dioxin-producing trash incinerators. It is exemplified by Love Canal, in Niagara Falls, New York, where local residents, who had been exposed to toxic chemicals, organized to demand the actions that finally led to the evacuation of the area and some compensation for their financial losses. Lois Gibbs, who led that fight, has formed a national federation of similar community groups. For such groups, the front line of the battle against chemical pollution is not in Washington but in their own communities. For them, the issues are clear-cut and are not readily compromised: A waste-management company decides to build a trash-burning incinerator, but the community, fearful of the health effects, doesn't want it. A chemical company keeps dumping toxic wastes in a leaky lagoon, but the community wants the practice stopped and the lagoon cleared up. In these battles, there is little room for compromise; the corporations are on one side and the people of the community on the other, challenging the corporation's exclusive power to make decisions that threaten the community's health.

Thus, confronted by powerful corporate opposition, the environmental movement has split in two. The older national environmental organizations, in their Washington offices, have taken the soft political road of negotiation, compromising with the corporations on the amount of pollution that is acceptable. The people living in the polluted communities have taken the hard political road of confrontation, demanding not that the dumping of hazardous waste be slowed down but that it be stopped; not that dioxin-producing incinerators be equipped with unworkable emission controls but that they be abandoned in favor of more benign recycling technology. The national organizations deal with the environmental disease by negotiating about the kind of Band-Aid to apply to it; the community groups deal with the disease by trying to prevent it.

It can be argued, of course, that, given the considerable power of the polluters, it is better to compromise with them, in the hope of making some partial improvement, than to engage them in a battle, which may only end in frustration. On the other hand, there is a danger that in the course of negotiating a compromise the environmental organizations will become hostage to the corporations' power and will experience the Stockholm Syndrome, in which hostages take on the ideology of their captors. Perhaps something like this happened to Jay D. Hair, the executive vice president of the National Wildlife Federation, which is heavily involved in "enlightened and responsible dialogue with corporate leaders." Hair recently admonished environmentalists that "our arguments must translate into profits, earnings, productivity, and economic incentives for industry"—precisely the arguments of corporate executives against the regulation of pollution.

Because the effort to improve the environment is so closely linked to the decisions that govern the technology of production, it is inevitably drawn into the realm of politics. Though technological choices are often thought of as being outside the realm of public policy or politics, and driven instead by "objective" scientific considerations, the evidence contradicts this view. That high-compression automobile engines are particularly powerful and generate smog is certainly an objective scientific fact. The decision to manufacture such an engine rather than one that is less powerful but does not produce smog is, however, not a scientific necessity but a human choice. When the auto companies decided to build high-compression engines, the choice, as we know, was motiviated by a private interest—higher profits. But their decision created a social problem—smog. When the decision was made, no questions were asked about the environmental effect of the new engines, and their social impact was revealed, too late, by the blanket of smog that covers the cities.

It is, of course, quite possible to choose a technology on the basis of its social consequences. When environmentalists speak of "appropriate technology," they have in mind production methods that not only enhance environmental quality but fulfill such other social purposes as conservation of energy and material resources and humane conditions of work. The social significance of a technological decision is much wider than its impact on environmental quality. The series of decisions that substituted plastics for leather, for example, aggravated not only environmental pollution but unemployment as well, for the plastics industry uses more capital and less labor than the leather industry did. In this sense, social guidance of technological decisions is vital not only for environmental quality but for nearly everything else that determines how people live: employment; working conditions; the cost of transportation, energy, food, and other necessities of life; and economic growth. And so there is an unbreakable link between the environmental issue and all the other troublesome political issues.

Efforts to develop a frankly political approach to the environmental issue—that is, to seek political power in order to use it to solve the environmental crisis—inevitably confront this linkage to the whole of politics. The most striking example is the brief but remarkable history of the Green Party in West Germany. In 1983, only four years after it was founded, the Green Party obtained 5.6 percent of the vote in the national election and won twenty-seven seats in the Bundestag. The Party began with an ecological outlook that emphasized not only the hazards of pollution but also the dangers of the ultimate ecological disaster—nuclear war. The Greens' antiwar position was an important source of votes, for neither the Social Democrats, who were then in power, nor the opposition Christian Democrats were against the expanding American nuclear presence in West Germany and the rest of Europe. In the 1983 election, peace-oriented voters, who were not necessarily preoccupied with ecological issues, found in the Greens a way to express this concern. This helped the Greens appeal to a group of voters considerably broader than ecological activists.

Yet the Greens' preoccupation with ecology was also a serious political drawback. The 1983 election occurred at a time of high unemployment, unprecedented in recent West German history. Some Greens believed that the best way to reduce environmental degradation was to lower the level of industrial activity, which they regarded as inherently antiecological, and this belief was translated into an open disregard for the demand for more jobs; after all, more jobs meant putting more people to work in polluting factories. Naturally, unemployed workers and many voters who sympathized with them were not attracted to the Greens.

Thus, from its inception the Green Party, like the environmental movement generally, ran head on into the firm bond linking ecology and economics. In the last few years, this confrontation has given rise to a basic split within the Party. One group consists of the "fundamentalists," who adhere to what they regard as basic ecological principles: advocacy of a social structure based on "unity with nature"; a spiritual devotion to harmonious relations among people and nations; respect for all forms of life. At a recent Green convention, one member of this group pleaded passionately for a resolution condemning the "slaughter" of thousands of frogs in biology laboratories, comparing it with the Holocaust. The fundamentalists are concerned chiefly with combating pollution and fostering a transition to renewable resources, "soft" alternative technologies, and a way of life appropriate to these. The other group consists of the "realists." They recognize the fundamental origins of ecological and other problems in the political economy. They are concerned not only with the technology of production and its impact on both the environment and jobs but also with which social class holds the governing power. The realists are linked to labor unions, and most of them favor political alliances—with the Social Democrats, for example, whose positions on environmental and peace issues have of late moved closer in some respects to the Greens' position.

In October of 1985, the split between the fundamentalists and the realists took a dramatic turn when, over the objections of the national

Green leadership, the realist-led Green Party of the industrial state of Hesse agreed to join the Social Democrats in a governing coalition. (The coalition broke up early in 1987 over a disagreement about a nuclear facility.) The Christian Democrats quickly attacked the Hesse Social Democrats for making a pact with "fanatical opponents of our free economy." In the heat of the accompanying debate, the fundamental political issue generated by ecological concerns became plain when a Christian Democrat who opposed the coalition declared that major corporations would leave Hesse, because "free companies decide where they want to invest." Thus, the history of the Green Party is a lively and effective dramatization of the links that tie ecology to economics and both to politics. It also illustrates the ease with which a failure to understand these links properly can lead to a politics that rises in defense of laboratory frogs but not unemployed workers.

Italy provides a different example of environmental politics. In the 1985 elections, environmentalists organized "green lists" of candidates, and some were elected to office. Although the Italian greens are strongly oriented toward environmental issues—in particular, opposition to nuclear power, to nuclear weapons, and to industrial pollution—most of them share the general political orientation of the Italian parties of the left. This leads to a kind of "red-green" situation, in which environmental concerns are joined to a broader left political program. In this sense, the green lists have also become important through the pressure they exert on the parties to adopt environmental positions; for example, recently they have helped to reverse the Italian Communist Party's earlier support for nuclear power. In turn, such developments may encourage environmentalists to work within the parties and thus link environmental concerns more firmly with the rest of politics.

In the United States, environmental politics has followed a distinctive course. Here the environmental movement is part of the phenomenon that since the Second World War has given rise to wave after wave of popular, issue-oriented movements: for civil rights, against nuclear-weapons testing; for women's rights; for gay and lesbian rights; against the war in Vietnam; for the environment; against nuclear power and for solar energy; for world peace. These movements have a good deal in common. All of them have arisen outside the arena of conventional politics, sparked by outsiders like Martin Luther King, Jr., and Rachel Carson rather than by established political figures. Their level of public support has typically gone through successive cycles of enthusiasm and apathy. At their height, some of these movements have achieved notable successes—the civil-rights and environmental laws, the nuclear-test-ban treaty, the new employment opportunities for women—all of them accomplished by non-electoral means: marches, demonstrations, lobbies. Yet, as the record of the Reagan Administration shows, these successes can be quickly eroded when officials hostile to them are elected to power. Indeed, the movements' greatest failure is their inability to translate the millions of votes that their combined adherents represent into significant electoral power, and put people in office who will protect their gains and expand them.

A major reason for the tenuous connection between the movements and politics is that their issues have been consigned to the political ghetto that is reserved for "special interests." In a sense, this isolation is self-imposed, for the varied concerns are usually manifested—in a legislator's office, for example—as single-minded constituents, one pleading for peace, another for sexual equality, a third for environmentalism, on down the list. Each special pleading demands a special response, at best unrelated to the other issues, and often in actual conflict with them. Each of the issues is regarded as a possible modifier of national policy but not as a creator of it—a correction in the course of the ship of state but not a motive force. Yet the issues that the movements represent, taken together, and added to those of the much older labor movement, constitute not only the major aspects of public policy but its most profound expression: human rights, the quality of life, health, jobs, peace, survival. What can unite these movements to enable them to exert an effect on national policy that expresses the profound political meaning of their collective issues?

The environmental experience suggests an answer. As we have seen, the obvious manifestations of the environmental problem—smog, toxic dumps, nuclear catastrophes—that set it apart as a "special interest" are but perceptible expressions of a deeper, underlying issue: how the choice of production technologies is to be determined. Here environmentalism reaches a common ground with all of the other movements, for each of them also bears a fundamental relationship to the choice of production technologies. Discrimination, for example, resembles pollution in that one of its major features—paying women and racial minorities less than white men—originates in decisions made by the managers of productive enterprises. Of course, other factors, social, cultural, and psychological, are involved, but the end result—wage discrimination—is, after all, an effective way of reducing production costs. Or, to look at this relationship from the other direction, if it were to be determined through some system of social governance that all the employees in an enterprise who do comparable work should receive equal pay, regardless of sex or race, a major effect of the social, cultural, and psychological forces than engender discrimination would be nullified. The connection between this common ground and the issues of peace and foreign policy is less direct but nevertheless substantial. It explains, for example, the apparent justification for the administration's military belligerence—the insistence that force must be used wherever it is needed to support governments and political groups that, like the administration, favor free-market principles of economic governance. In particular, it would be difficult to explain the administration's deep-seated antagonism to the Soviet Union—which engenders the monstrous threat of global nuclear destruction—if it were not for the opposite views of the two governments on the stewardship of economic enterprises.

Perhaps the most useful outcome of the environmental experience is that it illuminates the relationship between the outward manifestations of the ills that trouble modern society and the common origin of those ills. But there are risks in expounding this relationship. Calling attention to their source may appear to minimize the importance of the immediate problems that initially attracted adherents to the cause, and so risk their ire. At the same time, an effort to transform the "special interest" into a critique of basic and even more troubling faults in the social structure is likely to intensify opposition to it. The path taken by Martin Luther King, Jr., in the last few years of his life is a cautionary example. At the height of his influence, King had won major victories and had acquired a broad following as the leader of a powerful attack on outward expressions of legally enforced racial discrimination: segregated schools and public facilities. Then, a few years before he was assassinated, he began to link racial discrimination to its origins and, thereby, to other social issues. He sensed, it seems, that blacks could not break out of their persistent social ghetto if they remained trapped, as a "special interest," in a political ghetto. And so he declared his opposition to the war in Vietnam, led a march of poor people (black and white) on Washington, and championed the cause of striking black garbagemen in Memphis. In this new role, King quickly became more controversial. He was out of his depth, it was said—taken in by political radicals and diverted from his true mission. But he died believing, it seems, that this new course had brought him closer to the heart of the problem he had set out to resolve: that beneath the legal basis of racial discrimination lay the deeper problems of poverty and violence; that the root of racial discrimination is also the root of poverty and war.

Neither the environmental movement nor any of the other issue-oriented movements has yet attempted to follow the road that Dr. King began to travel before he died. But the environmental experience powerfully illuminates that road—a historic passage toward a democracy that can exert its force on the germinal decisions that determine whether we and the place we inhabit will thrive.

9

Environmental Pollution: High and Rising

Gus Speth

POLLUTION IS NOW OCCURRING on a vast and unprecedented scale around the globe. Trends, particularly since World War II, have been in two directions: toward large and growing releases of certain chemicals, principally from burning fossil fuels, that are significantly altering natural systems on a global scale, and toward steady increase in the use and release to the environment of innumerable biocidal products and toxic substances. These trends pose formidable challenges for societies, both industrial and developing—challenges that modern pollution control laws address only partially.

The dramatic changes in pollution in this century can be described in terms of four long-term trends.

From modest to huge quantities. The twentieth century has witnessed unprecedented growth in human population and economic activity. World population has increased more than threefold, gross world product by perhaps twentyfold, and fossil fuel use by more than tenfold. In the United States, gross national product has increased ninefold in this century, and fossil fuel use has more than doubled since 1950.

With these huge increases in population and economic activity have come huge changes in the quantity of pollutants released. Consider how increased fossil fuel use influences sulfur dioxide and nitrogen oxide emissions. These products of fossil fuel combustion are among the principal sources of smog and other urban pollution; they are also the pollutants that give rise to acid rain. Between 1900 and 1980, annual sulfur dioxide emissions grew by about 500 percent globally. In the United States, sulfur emissions grew by about 150 percent between 1900 and 1980, and nitrogen oxide emissions increased by about 900 percent.

From gross insults to microtoxicity. Before World War II, concern about air and water pollution focused primarily on smoke and sewers, problems with which people have grappled since the dawn of cities. The emergence of the chemical and nuclear industries fundamentally changed this focus on gross insults. Paralleling the dramatic growth in the volume of older pollutants, such as sulfur dioxide, has been the introduction in the post-World War II period of new chemicals and radioactive substances, many of which are toxic in even minute quantities and some of which persist and accumulate in biological systems or in the atmosphere.

Pesticides, one major product of the modern chemicals industry, are released into the environment precisely because they are toxic. Globally, pesticide sales have skyrocketed from about $5 billion in 1975 to $28 billion in 1985 (calculated in 1977 dollars). Projected sales for 1990 are $50 billion, a tenfold increase since 1975.

From First World to Third World. A myth easily exploded by a visit to many developing countries is that pollution is predominantly a problem of the highly industrialized countries. While it is true that the industrial countries account for the bulk of pollutants produced today, pollution is a serious problem in the developing world, and many of the most dramatic and alarming examples of its consequences can be found there.

Data from the United Nations Global Environmental Monitoring System indicate that, by and large, cities in Eastern Europe and the Third

World are consistently more polluted with sulfur dioxide and particulates than most of the cities in Organization for Economic Cooperation and Development (OECD) countries.

Third World populations also rank high in their exposure to toxic chemicals. In a sample of ten industrial and developing countries, three of the four countries with the highest blood lead levels of their populations were Mexico, India, and Peru; for the same ten countries, DDT contamination of human milk was highest in China, India, and Mexico. And what may be the worst industrial accident in history occurred in Bhopal, India, in 1984 when more than 2,000 people were killed. The accident occurred when a chemical used in the manufacture of pesticides, methyl isocyanate, escaped and drifted into crowded, low-income housing settlements adjoining the Union Carbide facility.

From local effects to global effects. When the volumes of pollution were much smaller and the pollutants were more similar to natural substances, the impact of these pollutants tended to be confined to limited geographic areas near their source. Today, the scale and intensity of pollution make its consequences truly global.

Nothing better illustrates the broadening of the concern about pollution from a local affair to a global one than the evolution of concern about air pollution. Atmospheric scientist John Firor's metaphor of the "endangered species of the atmosphere" is apt. Local air pollution is improving in some cities, but it is worsening in others, and it is hardly solved anywhere. Meanwhile, overall use of fossil fuels, and resulting emissions of traditional pollutants such as sulfur and nitrogen oxides, continues to climb. Acid rain, ozone, and other consequences of these pollutants are affecting plant and animal life over vast areas of the globe. Depletion of the stratosphere's ozone layer is a matter of such concern that an international treaty, negotiated to sharply reduce emissions of chlorofluorocarbons (CFCs), has been widely viewed as too weak even before it has been fully ratified. And, probably most serious of all, the buildup of greenhouse gases in the atmosphere—largely a consequence of the use of fossil fuels, CFCs, and various agricultural activities—continues, threatening societies with far-reaching climatic changes, a rise in sea level, and other consequences. These interrelated atmospheric issues probably constitute the most serious pollution threat in history.

With these long-term trends in mind, it is useful to examine what has happened in the last decade and a half since the United States put the spotlight on so-called traditional pollutants and initiated major, expensive cleanup programs. Are these programs achieving their goals? When we examine the available information, the pattern that emerges is one of some success, some backsliding, and a lot of holding-our-own. Judged by the goals of the Clean Air Act of 1970 and the Water Pollution Control Act of 1972, progress has been disappointingly slow. Also, more recently acknowledged problems, such as hazardous waste sites and groundwater pollution, for which correctional efforts have just begun, must be weighed against the gains that have been made.

On the positive side, the emergence of environmental concern in the 1960s and the cleanup programs that have followed in the 1970s have resulted in impressively large pollution abatement efforts. The United States spent about $739 billion (calculated in 1982 dollars) on pollution control between 1972 and 1984, with about 42 percent of this going to clean the air, 42 percent to water cleanup, and the rest primarily to solid waste disposal. About two-thirds of this has been spent by private business, with government and individuals accounting for the rest. Many manufacturing and other corporations have made serious and responsible efforts to achieve and sometimes exceed pollution control requirements. In the process, they have brought about major technological advances in pollution abatement and a pollution control industry in the United States that created an estimated 167,000 jobs in 1985.

A majority of dischargers have complied with discharge requirements without the need for legal enforcement actions, but in numerous cases, polluters have been recalcitrant, fighting every step of the way. Enforcement actions by state and local authorities are commonplace. Thousands of administrative enforcement orders are issued by the Environmental Protection Agency (EPA) every year, and hundreds of lawsuits against polluters are referred to the Justice Department annually for civil or criminal action.

What are the results of all the activity? On water pollution, the results have been mixed. Discharges of certain key pollutants have been reduced substantially, such as oxygen-demanding waste from industry and municipal sewage systems, which declined 71 percent and 46 percent, respectively, between 1976 and 1982. The quality of many important waterways, such as the Potomac River, has improved greatly.

If we examine the national picture as a whole, however, progress does not seem terribly impressive. A recent Conservation Foundation overview concluded that "in most cases little more has been done than to prevent further degradation." About two-thirds of U.S. surface waters have shown little overall change in water quality between 1972 and 1982. One important measure of water quality is the dissolved oxygen content of the water. Between 1974 and 1981, 17 percent of the U.S. monitoring sites reported improved oxygen levels, while 11 percent reported deteriorating trends, and the remaining 72 percent showed no significant change. Overall improvement was reported for bacterial and lead contamination, but nitrate, chloride, arsenic, and cadmium pollution showed significant increases. One problem is that, thus far, runoff and other nonpoint sources of pollution have escaped strict control.

It is possible to be more positive about air pollution. Between 1975 and 1984, U.S. emissions of particulates fell 33 percent, sulfur dioxide 16 percent, and carbon monoxide 14 percent. (The exception is nitrogen oxides, which showed little change between 1975 and 1984.) As a result of these gains, urban air quality in the United States, in general, has improved significantly.

These gains, while real enough, are modest when viewed against the environmental cleanup goals of the early 1970s. In every air pollutant category save one (particulates), total U.S. emissions today still exceed two-thirds of the 1970 amounts. The bulk of the pollution that gave rise to the Clean Air Act in 1970 continues. For example, total emissions of volatile organic compounds, a key ingredient in the making of urban smog and its principal component, ozone, were still 22 million metric tons in 1984, or 80 percent of 1970 emissions. The number of U.S. counties failing to meet clean air standards remains high. In 1985, 368 areas were not in compliance with the ozone standard (80 million people live in these noncomplying counties); 290 in noncompliance with particulate standards; 142 in noncompliance with carbon monoxide. In short, air pollution, like water pollution, remains a serious national problem.

These modest gains in controlling so-called traditional air and water pollutants look better when one compares them with what might have been. Between 1970 and 1985, U.S. GNP has grown by about 50 percent in real terms, and population has grown as well. Far from keeping pace with this growth, pollution actually declined. Overall, the U.S. economy generates much less pollution per dollar of GNP today than it did two decades ago. For example, between 1975 and 1985 U.S. electrical utilities reduced sulfur dioxide emissions from coal burning by 44 percent per unit of electricity generated, spending some $62 billion on sulfur dioxide controls during this period.

There are several important areas where environmental protection measures have had dramatic results. Lead emissions have taken a steep downturn as a result of federal regulations; these requirements had the effect of reducing lead use in gasoline by 78 percent between 1970 and 1985. Environmental releases of polychlorinated biphenyls (PCBs) have dropped off sharply as a result of a congressional ban of their manufacture, one of the provisions of the Toxic Substances Control Act of 1976. Beginning with the bans on DDT in 1972 and on Aldrin-Dieldrin in 1974, the EPA has prohibited most uses of organochloride pesticides. As a result of these decisive efforts, the presence of these toxic substances in the environment has shown impressive improvement. Between 1976 and 1980, blood lead levels in the United States declined sharply, while DDT concentrations in human tissue are only a fourth of what they were in 1972. The proportion of the U.S. population with PCB levels above one part per million declined from about 70 percent in 1972 to about 10 percent in 1983.

The presence of DDT, PCBs, and several other toxic substances has been greatly reduced because society has seen fit essentially to outlaw these substances and not merely to regulate their release. Industry has been required to innovate technologically and find substitute products, and not merely to use pollution-control devices. There is a lesson here.

As we face the future, it is worth recalling what antipollution efforts are up against. The recent report of the World Commission on Environment and Development, *Our Common Future,* stated the situation as follows:

The planet is passing through a period of dramatic growth and fun-

176

damental change. Our human world of 5 billion must make room in a finite environment for another human world. The pollution could stabilize between 8 billion and 14 billion people sometime in the next century, according to UN projections. . . . Economic activity has multiplied to create a $13 trillion world economy, and this could grow five or tenfold in the coming half-century.

These challenges will require far-reaching responses. Certainly, we must continue and strengthen the efforts already begun. The regulatory programs of the industrial countries have yielded definite results over the past two decades, particularly when judged against the economic expansion of this period, and continuing challenges will require that these programs be enhanced and improved. Monitoring and enforcement capabilities must be strengthened; new types and sources of pollution must be tackled; intermedia effects must be attended to; and the overall regulatory process must become more cost-effective, efficient, and streamlined as demands mount for those on both sides of the bargaining table. The need for developing countries to face up to their escalating pollution problems is also acute.

Yet, it is certain that more of the same, even if better, will not be enough to cope with the pollution challenges identified here. More fundamental changes will be needed.

From its origins in the early 1970s, U.S. air and water pollution control legislation has recognized that tighter standards could be applied to "new sources" of pollution, in contrast to existing plants, because new sources present the opportunity to go beyond end-of-pipe removal of waste products and to build in process changes that reduce or eliminate the waste that must otherwise be removed. This basic concept—source reduction instead of emissions control—*writ large,* is fundamental to solving world pollution problems. In the long run, the only affordable and effective way to control pollution of the scale and type societies confront today is to work against the tide to change the products, technologies, policies, and pressures that generate waste and give rise to pollution.

To implement this requirement, and to deal with the serious pollution challenge in the decades ahead, several large-scale social and technological transitions are needed. The pollution societies confront today is a huge, multifaceted phenomenon that cuts across economic sectors and geo-

177

graphical regions. It is integrally related to economic production, modern technology, life styles, the sizes of the human and animal populations, and a host of other factors. Pollution is likely to yield only to broad social transitions that have multiple benefits.

Transition one: the shift away from the era of growing fossil fuel use toward an era of energy efficiency and renewable energy. The extraction, transport, and particularly the combustion of fossil fuels create a huge portion of the world's environmental pollution, including major contributions to acid rain, forest death, and the greenhouse effect.

Transition two: the move away from an era of capital- and materials-intensive, "high-throughput" technologies to an era of new "closed" technologies. These new technologies use raw materials with great efficiency, rely on inputs that have low environmental costs, and recover and recycle materials—thus generating less pollution.

Transition three: the change to a future in which societies actually apply our best science to design with nature. Much pollution regulation, for example, has been based on a one-pollutant-one-effect approach, often with the requirement that tight causal connections to environmental damage be identified for the particular pollutant. That, however, is not the way biological systems work. When we look at these systems, human or ecological, we see that they are subject to multiple and interacting stresses, chemical and physical, natural and manmade. New regulatory concepts are needed.

Agro-ecology offers an alternative to industrial agriculture, one of the leading sources of pollution today. The ecological approach to agricultural production stresses low inputs of commercial fertilizers, pesticides, and energy; the biological recycling of energy and nutrients; and primary reliance on naturally occurring methods for crop protection. Combined with the rapid progress being made in organic farming and integrated pest management, an alternative to the exclusive use of pesticides, agro-ecology offers the prospect of an agriculture redesigned to be sustainable both economically and environmentally.

Transition four: a move to an environmentally honest economy in which policies do not subsidize raw materials or the generation of waste; in which private companies and governments "internalize the externalities" so that the prices of goods and services in the marketplace reflect the true social costs of production, including the costs of pollution to society; and where national income accounts treat the depreciation of natural assets just as rigorously as capital assets. Creating efficient economic incentives and disincentives is essential to long-term pollution control, and making the market mechanism work better by getting the price right is an essential means to that end.

Transition five: the move to more international approaches to reducing pollution. As we have seen, pollution problems are increasingly international and even global; international solutions must become far more common. Whether one likes it or not, the diplomatic agendas of nations increasingly will feature pollution and other environmental issues.

Transition six: a demographic transition to a stable world population. Through appropriate policies promoting economic advancement, education, health, the status of women, infant care, and family planning services, global population growth should be halted before it doubles again. It will be difficult enough to cope with pollution problems in a world in which population has grown from 5 billion to 8 billion people.

These six transitions are the product of a deep appreciation of the importance of economic and technological forces in the modern world. But if solutions are found, they will come from another realm as well, from the hopes and fears of people, from their aspirations for their children and their wonder at the natural world, from their own self-respect and their dogged insistence that some things that seem very wrong are just that. People everywhere are offended by pollution. They sense intuitively that we have pressed beyond limits we should not have exceeded. They want to clean up the world, make it a better place, and be good trustees of the earth for future generations. Like Thoreau, they know that heaven is under our feet as well as over our heads. Politicians around the globe are increasingly hearing the demand that things be set right. And that is very good news indeed.

10

Environmentalists at Law

Fredric P. Sutherland
and Vawter Parker

DAVID SIVE, the New York attorney who twenty years ago brought some of the country's first environmental lawsuits, has said that "in no other political and social movement has litigation played such an important and dominant role. Not even close." This is observed fact, not opinion. In the last two decades:

- Lawsuits saved millions of acres of land from inappropriate development, sometimes with injunctions that halted bulldozers and chain saws virtually at the last moment.
- Litigation protected hundreds of species of plants and animals threatened with extinction, both by preserving essential habitat and by eliminating specific threats to their existence.
- Lawsuits helped improve air and water quality in America. More important, these suits prevented pollution from worsening.

- Litigation put teeth in the laws enacted by Congress and the state legislatures to protect the environment and made the environmental movement credible.
- Litigation provided a high-leverage means for citizen activists to confront successfully the sometimes overwhelming resources of government and industry.
- Litigation was widely reported in the news media and attracted support for environmental causes.
- Litigation established important precedents and new legal principles that changed the rules by which environmental decisions are made.

But all that is in the past, and the question is, will environmental litigation be necessary in the future? The short answer is, yes, it will. Indeed, it is to be hoped that there will be a great deal more of it.

The environmental movement is based on law as much as it is based on information and education. However idealistic many environmentalists may be as individuals, their movement is anything but utopian. At its core is the assumption that most people and institutions, even the most well meaning, will try to maximize their own benefits while transferring their costs to someone, or somewhere, else. The fundamental parable of the environmental movement is the tragedy of the commons, summarized by Garrett Hardin as "freedom in a commons brings ruin to all." Walt Kelly's Pogo recognized that we all share this trait when he reported, "We has met the enemy, and it is us."

It is not surprising, therefore, that those conservationists who were not in the courtrooms in the 1960s and 1970s were in the offices of their senators, representatives, and state legislators or at home writing them letters. And they got results. Without attempting a count, probably a thousand statutes in the federal and state codes exist today that are principally intended to protect the environment.

We now see that statutes are not enough. However good their intent and however competently they are drafted, statutes are not self-executing; they have to be implemented. More to the point, *someone* has to implement them. Regardless of what we were told in high school civics class, the executive branch does not necessarily carry out the laws that Congress passes. There are a lot of reasons why what should happen often does not.

Most environmental statutes require some government agency to implement them. Some agencies truly may not have enough money or enough employees to do so.

Some agencies are filled with people who have spent their entire careers as integral parts of the problem. It is probably too much to expect that these people will or even can retool their minds and spirits just ten years short of retirement to become functioning parts of the solution. (The Bureau of Reclamation, which recently announced that it is going out of the dam-building business and will become a water conservation agency, comes immediately to mind, but only barely noses out the Federal Regulatory Commission.)

Some agencies are psychological captives of the predators they are supposed to guard against (the Forest Service with regard to the timber industry, the Bureau of Land Management with respect to the cattle industry). Almost none can say no to a senator or representative who does not want the law applied to a major contributor or to too many of his constituents. And, of course, government agencies are no more immune than any other institution to occasional incompetence or even venality.

The upshot is that agencies that are supposed to carry out environmental statutes very often delay, miss deadlines, convert mandatory standards to discretionary ones, create loopholes, water down strict statutes in the regulatory process, or simply refuse to use their enforcement powers when faced with blatant violations.

This comes as no surprise. After all, the executive branch of the federal government is itself one of the greatest engines of environmental destruction ever loosed upon the land. Conservationists have long been aware of this, and the very first statute of the so-called environmental decade—the National Environmental Policy Act, which became law on January 1, 1970—was aimed not at private individuals or industry but at the federal government.

The poor performance of the executive branch should make us mad but should not discourage us. One cannot deny that real progress has been made in most of the areas in which Congress and the states have legislated, or at least that greater potential for progress now exists. The fact is that law, which is slow, almost always slower than we would like, and which sometimes gets off the track, is the only instrument available to us other than education—which is even slower, and with some people never takes at all.

Ultimately, all we can do is try to make sure that the laws address the right problems and are carried out as effectively as possible. Environmental protection, like liberty, requires eternal vigilance.

Hence, the necessity for, and the growing desirability of, environmental lawyers and environmental litigation. Without environmental lawyers representing concerned citizens willing to go to court, much of the legislation that Congress has passed and the President has signed would be meaningless. Under some statutes lawyers and citizen groups may be able to enforce the law directly: In the Clean Water Act and the Resource Conservation and Recovery Act, for instance, Congress has attempted to provide for citizen enforcement against some polluters.

To date, however, environmental lawyers have made their greatest contribution not in going after corporations and individuals who have failed to comply with environmental laws, but in bringing suits to compel government agencies and officials charged with administering environmental laws to do their jobs. The most common example is the preparation of environmental impact statements in which an agency must acknowledge the adverse as well as beneficial impacts of its proposed action and outline alternative actions available to it; seemingly countless suits were necessary in the early years of the National Environmental Policy Act to get agencies to comply at all.

There are many other sins of omission. For instance, the U.S. Fish and Wildlife Service (and, in the case of marine mammals, the National Marine Fisheries Service) must officially list endangered species and designate critical habitat for them before the sanctions of the Endangered Species Act apply; both agencies have tried to shirk that responsibility rather than risk irritating commercial interests, politicians, and other federal agencies.

The Forest Service must adopt management plans for individual national forests before the environmental constraints those plans are supposed to contain can apply to individual logging contracts; the Forest Service has managed to delay adopting such plans for years.

The Army Corps of Engineers must assert its permitting jurisdiction over placer mining operations in Alaska's rivers before standards to protect fish and water quality apply. Only under the threat of a lawsuit is it finally doing so.

The Environmental Protection Agency (EPA) must issue standards for a pollutant before the government or anyone else can enforce those standards. Dozens of lawsuits have been necessary, and dozens more will be necessary, just to force EPA to issue the regulations under the Clean Air Act. Dozens more will be necessary to make the regulations tough enough to do the job.

Litigation works, for both institutional and psychological reasons. If it is unpleasant for an individual or an agency to comply with the law, a court order can make it a lot more uncomfortable not to comply. It can prohibit an individual or agency from taking any action, including action that it wants to take, until it complies with the law. It can freeze funds, the lifeblood of any agency or action. It can effectively isolate the agency or individuals within it by forbidding others to cooperate in the illegal action or failure to act, on pain of contempt of court proceedings.

Consider the injunction: a simple court order, but an order enforceable by contempt proceedings, if necessary, and one of the most enlightened and effective instruments of government ever devised. Injunctions have stopped developers from building dams, highways, power plants, and logging roads. They have stopped the government from licensing dangerous chemicals. And they have stopped the destruction of forests, rivers, and wilderness areas.

But court orders are not always negative—they can force positive change as well. For example, in 1977 Congress directed EPA to investigate whether airborne radionuclides should be regulated as a hazardous air pollutant. Airborne radionuclides emanate from a variety of sources, including uranium mill tailings, high-level radioactive wastes, phosphorous plants, and nuclear fuel facilities. The amount and type of radioactivity varies from source to source, but one thing is clear—none of it is good for the public's health. The evidence is overwhelming that radionuclides cause cancers and genetic defects; hence, Congress's concern.

In 1979, EPA agreed that airborne radionuclides were indeed hazardous and should be controlled. Under the Clean Air Act, the agency then had 180 days to issue proposed emission control standards and an additional 180 days to promulgate final standards. The time limits passed without the required standards. When it became apparent that EPA had no intention of issuing them at all, conservationists filed suit in a federal court in California against the agency and its administrator.

The only real issue in the lawsuit was how much more time EPA should have to prepare the standards, given that the standards were already years overdue. When it became clear that the agency wanted additional years to prepare standards for some sources, the judge entered an order requiring that regulations be drafted within 180 days. He assumed that the final standards would be forthcoming 180 days after that, as required by the Clear Air Act.

Unfortunately, EPA still balked. Although the agency issued proposed standards, it refused to promulgate them in final, enforceable form. As a result, the conservationists went back to court. In July 1984, EPA was ordered to issue the final standards by October of that year. In September, the agency returned to court to ask for more time, a request the judge denied.

> I think it's outrageous . . . and I hope that somebody told Mr. Ruck-elshaus that.. . . [H]e's a very fine man . . . and I think the world of him. But since he's taken over that agency, he hasn't done much with radionuclides. And the United States Government is about five years late. And I'm not just about to give it any more time. I'm kind of burned up.

One would think that would have taken care of the matter, but it did not. EPA did not comply with the court's order, and the conservationists who brought the case were left with no choice but to ask the court to hold Ruckelshaus in contempt. Only then did EPA comply. Today, we do have emissions limits for airborne radioactivity. Slowly, but surely, litigation achieved a major public health goal.

Although environmental litigation will continue to be essential in years to come, there are troubling signs ahead. Progress is by no means inevitable. In addition to the anticipated increasing cost and complexity of such litigation are some special problems worth noting.

By the time he leaves office, President Reagan will have appointed about half the federal judges. Attorney General Meese and others have gone to unusual efforts to ensure that the president nominate only judges who agreed with them on a number of very specific points, and many of these nominees appear to be particularly hostile to environmental concerns. This does not bode well for future environmental success in the courts. Skilled advocacy will be more important than ever.

The United States Supreme Court is in itself a cause for concern. In the eighteen years since the enactment of the National Environmental Policy Act, the Supreme Court has decided ten cases brought by conservationists against the federal government for violations of the Act. Not once has it decided a case against a government agency. Professor Daniel Farber examined each of the ten cases and concluded,

> Almost all the Supreme Court decisions were unanimous—and the few brief concurring or dissenting opinions are rather lackadaisical, as if the dissenting justices barely thought it was worth the effort of expressing their views. . . . It is hard to read the Court's NEPA decisions without getting an impression of judicial indifference, if not distaste.

And that is not all. In the fall of 1987, the Supreme Court all but eviscerated the citizen's suit provisions of the Clean Water Act. In a particularly badly reasoned and badly written opinion that ignores the government's own dismal enforcement record, the court held in *Chesapeake Bay Foundation* v. *Gwaltney* that citizens cannot sue a polluter for violations of the Clean Water Act unless they can prove that the polluter was still violating the law at the time the suit was filed or that the polluter was likely to engage in illegal activity in the future.

The holding in *Gwaltney* in effect immunizes polluters from liability under the Clean Water Act, no matter how serious their violations. Under the court's reading, an industry can avoid any penalties simply by complying whenever it receives a citizen's notice of intent to sue, which the citizen must give the polluter at least sixty days prior to filing suit. Without liability for past violations, and with a sixty-day notice requirement, there is simply no incentive whatever to stop polluting. The court either does not understand the statutory scheme of the Clean Water Act or it is, through judicial interpretation, deliberately repealing the statute's most effective enforcement provision. Either way, the Supreme Court has now gone a long way toward putting citizens out of the enforcement business.

How striking is the contrast between the unanimous opinion of the Supreme Court in the *Gwaltney* decision and Justice William O. Douglas's dissenting opinion in *Sierra Club* v. *Morton* in 1972—which, though a dissent, seemed to hold promise for the future and possibly even herald a role for the Supreme Court in environmental cases comparable to its earlier leadership in the civil rights movement. In arguing that inanimate objects threatened with despoliation should be represented in court, Justice Douglas wrote,

> The voice of the inanimate object . . . should not be stilled.. . . [B]efore these priceless bits of Americana (such as a valley, an alpine meadow, a river, or a lake) are forever lost or are so transformed as to be reduced to the eventual rubble of our urban environment, the voice of the existing beneficiaries of these environmental wonders should be heard.
>
> Those who hike the Appalachian Trail . . . or run the Allagash in Maine, or climb the Guadalupes in West Texas, or who canoe and portage the Quetico Superior in Minnesota, certainly should have standing to defend those natural wonders before courts or agencies, though they live 3,000 miles away.

Nevertheless, conservationists in the last two decades have had little help from the Supreme Court, none of whose members appear to be particularly concerned about the environment.

Finally, one other development should be mentioned. Recently, a whole cottage industry has sprung up across the country under the name of alternative dispute resolution, or mediation. The idea, of course, is that if someone brings together the disputing parties in a conflict they may be able to resolve their differences without the costs and ill feeling of litigation. There are plenty of conflicts that should be resolved in just such a manner. Alternative dispute resolution, however, is hardly the panacea some would like to believe.

First and most important, true environmental disputes—those that are more than an argument over who has to have the trash dump or powerline in his back yard—almost always involve public resources and public policies. That is why Congress or the state legislature got involved in the first place, and that is why we have public officials, called judges, to decide particular disputes. Unlike private mediators, judges must take into account not only the desires of the opposing parties but also the public's interest as expressed in law. The public's interest is not served if the parties agree to resolve their dispute by ignoring the law. As the former editor of *Not Man Apart,* Tom Turner, has written,

> If Congress has passed a statute to protect water quality, for example, are private groups to decide that it should apply in some situations but not in others? Or to some industries but not to others? These disputes are supposed to be resolved by Congress and the courts. Private mediation may be fine when two businesses are haggling over which is responsible for a cost overrun, but it is Congress and the courts, the latter interpreting written law, that are supposed to decide larger societal issues.

Second, successful mediation requires that the opposing parties have comparable, though not necessarily equal, bargaining strength. There can be no meaningful compromise if one side has all the cards. Winning litigation—enforcing the law—can help equalize positions so that truly meaningful negotiations can take place.

Finally, the real attraction of alternative dispute resolution to some is the assumption that it will allow business to go on as usual, with minor measures taken to mitigate the environmental harm. Most conservationists, however, are not content just to preserve façades of buildings and zoo specimens of endangered species. And sunken, landscaped, or otherwise mitigated freeways still produce air pollution and gridlock. At the heart of many of the most important conservation disputes is a desire to change the way society operates, and the long-term conservationist cause may be better served at times by risking the loss of everything in litigation than by compromising. Supporters of the conservation movement will do their cause harm if they unthinkingly assume that alternative dispute resolution is simply a cheaper and more palatable substitute for litigation and fail to discern when one or the other is their best choice.

Despite these cautions, there is no fundamental reason to believe that litigation will, or should, be a less important conservation tool in the future. Slow it may be, underfinanced, most certainly, but litigation to enforce statutes and regulations will remain one of the most powerful tools for change that conservationists have. More of it will be needed to meet the challenges ahead.

11

Environmental Law— Twenty Years Later

J. William Futrell

ENVIRONMENTAL LAW WAS BORN with Earth Day and shared the hopes of that era. The environmental movement's efforts resulted in a massive legal response aimed at curbing air and water pollution and safeguarding threatened natural resources. The law's goals—like those of the movement—were high: nondegradation of air, rivers clean enough to swim and fish in, and a national environmental policy that would encourage harmony between man and his environment. With the hindsight of the first twenty years of environmental law, American environmentalists can take pride in the scope and intensity of their efforts, and satisfaction from significant successes in curbing major air pollutants and in cleaning up miles of riverways. But they also must acknowledge just how far they have fallen short in achieving their Earth Day objectives. This shortfall in environmental protection is paralleled by defects in the development of environmental law and administration.

The defects in environmental law are caused primarily by a distorted emphasis on administrative agency law, with its command and control regulation, and a failure to develop the environmental dimensions of tort, criminal, business, and planning law. The defects in environmental administration reflect the uncoordinated efforts of our Balkanized government structure, which divides responsibility, thereby impeding strategic leadership. What is needed most in environmental law is a renewed effort to build consensus in the political branches and a revitalized pluralism to expand the base support for environmental protection.

During the last twenty years, environmental law has gone from practical nonexistence to a highly developed field of law. In 1971, the Environmental Law Institute published a summary of environmental law that comprised thirty-three pages in the *Environmental Law Reporter*. In 1988, the text of the statutes alone takes up more than 800 pages. Since its inception, the *Environmental Law Reporter* has published more than 4,000 federal court decisions. These decisions are paralleled by many thousands of administrative decisions, which are equally important to lawyers seeking to advise clients.

Most of these actions took place at the federal level. Responding to strong public sentiment, Congress set national standards and pushed aside state regulatory programs deemed ineffective. The new federal statutes focused on command and control regulation, requiring extensive permitting, and establishment of a national bureaucracy. The Environmental Protection Agency, created by an executive order, is one of the few federal agencies without a congressional charter. Its authority over environmental insults is shared with eleven other federal departments and four other agencies. Congress has never moved on to create the federal Department of Natural Resources needed to pull these scattered efforts together.

The new environmental law was—and still is—intensely political. Congress has forty separate committees and subcommittees that deal with the environment, and environmental bills have to run the gauntlet of these congressional checkpoints. Unfortunately, each of the environmental statutes reflects different compromises. The result is a hodgepodge of overly detailed laws with sometimes conflicting demands. Congress often treats environmental bills the same way it treats the tax code. Not trusting the Internal Revenue Service, Congress tries to dictate the regulations of the

IRS by writing overly detailed statutes. EPA is in a similar situation.

Any evaluation of our performance needs to assess the progress as well as the failures. In the areas of law and administration, where Congress and EPA have coordinated their efforts, the United States has achieved considerable progress. Monitoring studies show significant decreases in the areas targeted for action by the 1970 Clean Air Act and the 1972 Clean Water Act. However, Congress and EPA ducked the issue on a whole range of vital issues. These two bodies have concentrated their attention on the easy, law-abiding industries and have ignored the scofflaw industries such as mining, which continues to thumb its nose even at the halfhearted compromises of the 1977 Surface Mining Law. (Some estimate that mining wastes do more damage to water quality than all of the industrial effluent and hazardous waste sites combined.) The coverage of environmental law resembles a checkerboard—with some activities regulated and others ignored.

The failure to address the problems of mining waste, municipal scofflaws, and nonpoint pollution from sources such as pesticides and road salt run-off is a major reason why America is not meeting its Earth Day goals in water pollution cleanup. Progress is being made in the areas marked for attention in the 1972 water law but not in the areas beyond law's empire.

In an effort to assess the efficacy of environmental law, we should note that Congress and environmental lobbyists have relied almost exclusively on command and control regulation administered by a large national bureaucracy. This approach is a wise one—command and control regulation is needed—but it does not go far enough. The country needs a strong EPA, but a regulatory approach alone will not suffice.

Environmental law is distorted. For the law to be effective in protecting the commons, environmental reformers need to focus on developing new legal strategies to aid their efforts. These strategies should include development of tort and criminal law and changes in corporate and planning law to make the market system work. Some of the building blocks are already on the books to create these new developments. What is needed is effort and commitment to make them happen.

Tort law is an important tool in ensuring health and safety in a whole range of activities. Regulatory agencies such as state medical boards, which certify doctors, and the Federal Aviation Administration, which certifies airline pilots, play an important role in protecting health and safety. But a larger disciplinary role is played by the possibility of lawsuits for damages. The fear of liability is a vital element in ensuring better performance in these and hundreds of other enterprises. The Superfund law represented the first major effort to graft liability law onto environmental law. Superfund rests on the principle that the polluter pays. The government can either force the polluter to clean up or pay for the cleanup itself and then sue the polluter for the expenses.

The liability provisions of Superfund are the subject of a heated debate now because cleanup is going slowly while the preliminary legal motions are handled in giant lawsuits. But the liability provisions are already a major success in environmental protection because manufacturers are improving their current waste-handling practices radically and are sharply decreasing manufacture of future wastes. This improvement in current and future practices is motivated by the liability provisions. For real advances toward Earth Day goals, greater use of liability law is needed to curb environmental insults.

Environmentalists and Congress need to get serious about criminal law. It is incredible to note that between 1881 and 1981 only fifteen environmental crimes had been prosecuted. And these cases involved only misdemeanors because no federal environmental statute provided felony sanctions. That picture is changing. Since 1982, prosecutors have obtained 384 indictments and gotten 291 convictions (84 corporations and 207 individuals). The trend is up and is now visible in state courts as well. Failure to use the criminal law during the 1970s and early 1980s signaled a lack of serious intent on the part of environmental protection leaders. To make progress toward Earth Day goals, criminal law must work in tandem with command and control regulation and liability law.

A major effort is needed to shape corporate law so that the market will reward environmental successes and punish environmental failures. A major reason for the 1988 lag in hazardous waste cleanup is that the insurance industry has withdrawn from the hazardous waste cleanup market. The old insurance policies governing environmental liability simply tried to build a wall between the insurers and the industrial assured, denying coverage for pollution-related activity. The policies did not have the policing effect that insurance has in other areas. Insurance leaders are now planning their eventual return to the market. When insurance comes back to the field of environmental coverage, it will apply risk assessment and rewards for risk management as it does in other fields. New and more sophisticated tools of environmental auditing will be required of assureds. This development may do more for cleanup than the whole first decade of EPA regulations.

More also needs to be done in the field of environmental administration. For the last twenty years we have muddled through with a system of divided environmental administration. Nobody would try to run a business or manage a volunteer campaign the way the United States runs its environmental administration. It is set up to magnify conflict and to increase the chances of gridlock.

Environmental programs are scattered throughout the government. Water pollution from strip mines is regulated by the Department of Interior, water pollution from chemical plants by EPA, and water pollution from road salts not at all. Construction in sensitive wetlands areas is regulated on a push-me-pull-you basis by both the EPA and the Army Corps of Engineers. Different standards for the same hazardous chemical are promulgated by the Occupational Safety and Health Administration and EPA.

This system is designed to fail because it avoids defining national goals and setting priorities. Our Balkanized environmental departments have lacked a sense of strategy as well as leaders who press to win and to reward those who champion environmental goals. We are lucky to have accomplished as much as we have in curbing criteria air pollutants and controlling some water pollutants.

Potential policy failure is built into U.S. government by our Constitution, which divides power horizontally between the legislative and executive branches of government and vertically between state and federal officials. For policy to succeed, these separations and divisions have to be overcome by leaders who help define objectives and forge a consensus in the public interest. This requires leaders with a sense of strategy—and the environmental arena has lacked these.

Since 1981, strategic environmental leadership has been especially difficult because of the low priority given environmental matters by the

White House. In our constitutional system it takes two to tango, and the White House seems to want to sit out the environmental dances.

But these problems of divided environmental leadership did not start with Ronald Reagan's inauguration. The same complaints have been voiced in each administration. However, in other controversial areas of policy, such as foreign aid and tax reform, Congress and the White House recognize the perils of divided command and forge compromise and consensus. It is past time to do that in environmental affairs if we are to address environmental problems further.

Soundly designed planning laws could ease these difficulties. One such law, the National Environmental Policy Act (NEPA), as suggested by its title and statement of goals, could serve as a lodestar in defining national environmental priorities. Unfortunately, NEPA's goals have largely been ignored by successive administrations, both Democratic and Republican, in drafting their budgets and staffing programs for the environmental agencies.

Court decisions have converted NEPA from a national environmental policy act into a national environmental procedures act. Currently, the law is useful in aiding intergovernmental coordination but not in helping us define a national strategy for environmental protection.

This will happen only if the national and grassroots environmental organizations renew their commitment to expanded citizen participation in the NEPA process. What NEPA planning law needs—as do other areas of environmental law—is more life, more citizen participation. Law reform can open up the process, but it cannot guarantee the result. Law is no substitute for politics.

While the federal government may have only marked time in a number of areas, the 1980s have been an explosive decade of action for environmental professionals in other sectors. During the 1980s state environmental programs have blossomed; state and local enforcement programs have become reality. For example, the New Jersey Department of Environmental Quality grew slowly through the 1970s. During the 1980s, the department doubled, and by 1988, with a staff of 3,400, the New Jersey department is a major national force in its own right. Its growth in staff, budget, and capability is representative of how state environmental programs become a reality in the 1980s. It is estimated that at the end of that decade 90 percent of all environmental enforcement actions take place at the state level.

In industry, the number and quality of hands-on environmental managers increased dramatically during the late 1980s. The development of these proactive, problem-solving programs in the business sector is one of the slow-building, silent success stories of the environmental field of the 1980s. Environmental auditing, only a vague concept in 1981, had become a standard practice by the end of the decade in almost all smaller companies.

The ranks of environmental professionals swelled during the 1980s. These new players present an opportunity to create an expanded base of support for more effective environmental protection programs.

It is easy to detail the failings of environmental law and where it has fallen short. The failures of environmental law underscore how complex the challenges are. The solution is not less law, but more law. Next to personal example, law is the most powerful teaching tool in society. Environmental professionals (lawyers and others) are more a part of the solution than a part of the problem. They are the problem solvers in bridging the gap between current stalemates and the realization of Earth Day goals.

12

The Earth's Environment: A Legacy in Jeopardy

Jay D. Hair

THE LONG ANTICIPATED year 2000, the subject of so much speculation and study, is less than twelve years away. The year will launch the twenty-first century; it will also start the third millennium A.D., a chronological juncture when society should ask, Will an era of global peace, human prosperity, and environmental quality ensue?

A millennium is defined not only as a period of 1,000 years, but also as a period of great happiness, peace, and prosperity. Whether the third millennium portends all that for the peoples of the world will depend greatly on the leaders and the ideals that we nurture today.

The leaders of the twenty-first century, the high school graduating class of the year 2000, enter kindergarten in the fall of 1988. The majority of children in the industrialized world will to go to school from homes where the larders are full, medical attention is consistent, and physical shelter is assured. But too many of the world's children will leave homes that are mired in poverty and disease to enter school for only a few years and then to be called back home to help eke out a subsistence livelihood. What kind of world will these vastly different people inherit? Equally important, what kind of leaders will our world inherit in the year 2000?

The complexities and interrelationships of modern society require that we take a global view as we look to the future. As Peter Raven, director of the Missouri Botanical Gardens, has eloquently stated,

> The world that provides our evolutionary and ecological context is in trouble—trouble serious enough to demand our urgent attention. The large-scale problems of overpopulation and overdevelopment are eradicating the lands and organisms that sustain life on this planet. If we can solve these problems, we can lay the foundation for peace and prosperity in the future. By ignoring these issues, drifting passively while attending to what seem more urgent, personal priorities, we are courting disaster.

In our world of complexities and contradictions, the problems are equally complex and contradictory. We live in a world in which our mastery of technology has enabled us to walk on the moon, replace a diseased heart with a fully functioning human one, store a million pieces of information on a computer chip smaller than a human fingernail, and even create human life in a Petri dish.

Yet, it is also a world in which technology has degraded natural resources and changed forever the wild places of the world; a world in which the chemical revolution has eased the daily chores of those in developed countries, while exposing every human being to potentially dangerous chemicals from the moment of conception to the time of death.

We live in a world of opulence, a world where a single painting by van Gogh sells for $40 million and as many as 100,000 people (40 percent of them children) starve to death each day. The World Bank has estimated that 800 million people live in conditions of absolute poverty, life degraded by disease, illiteracy, malnutrition, and squalor.

Modern society's technological miracles also have created modern society's most dreadful tools. The proliferation of nuclear arms has given man the unthinkable ability to destroy all life on earth. But instead of resolving to eliminate such gruesome devices, the United States is poised to develop a defense program that will take the arms race to a new and potentially more deadly dimension.

The realities of the world are harsh. Is it any wonder, then, that many people stand ready to mislead the public by pretending that the world's problems are not really severe; that environmentalists are elitists, alarmists, and against progress; that technology can solve most problems; that inactions and lack of government involvement prevent viable solutions; and that the developed world should be allowed to continue its selfish life style? This is the state of the world that the children of the twenty-first century will face.

How do concerned people begin to reconcile the world's contradictions, recasting them into opportunities for greater freedom? How do we bring the world back into harmony with nature without sacrificing the technical gains that have extended the average human life span? And how do we ensure both environmental quality and sustainable economic development in a world desperately in need of both?

While there are no simple answers, five broad imperatives should guide decision making in the decades to come.

First, society must acknowledge that human health is directly dependent on the health of our global environment. As we deplete our natural resources, as we continue to permit the extinction of species, as we continue to pollute our air and water with toxic substances, we are imperiling the quality of all life on earth. Only when we appreciate those facts fully and realize that they affect us personally can we hope to establish and bequeath a sustainable society.

Second, we must find new and creative ways to develop policies that guide our public and private institutions. We must integrate such varied disciplines as science, engineering, economics, and law as we seek solutions to the wide range of complex issues before society.

For too long we have focused our intellectual energy on single-discipline analysis, ignoring how one discipline affects and is affected by another. We can no longer afford such narrow thinking. We can no longer exclude or exempt certain segments of society from responsibility for environmental quality. Instead, we must, for instance, provide incentives to encourage the business community to apply its expertise in product development and marketing to resource conservation and environmental protection. At the same time, we must eliminate the economic incentives that encourage corporations to cut environmental corners in order to improve year-end profits. Consider one of the most costly examples: the disposal of toxic wastes.

Thousands of toxic waste dumps exist in the United States because, in the past, dumping or burying hazardous wastes was considered cheaper than proper disposal. But cheaper for whom?

Today, U.S. taxpayers are spending billions of dollars to clean up abandoned waste sites that never would have been created if the free enterprise system had built-in incentives for long-term environmental protection. It does not. Indeed, in many cases, the corporate environment encourages just the opposite.

Washington State University researchers, for instance, recently looked at the ethical codes of 200 of the *Fortune* 500 companies. They found that 75 percent of them fail to address the firm's role in civic and community affairs. Moreover, three-quarters of the codes fail *even to mention* the moral imperative of environmental safety.

The May 9, 1988, issue of *Newsweek* magazine noted that "with insider trading damaging Wall Street and fiscal improprieties plaguing several big companies in recent years, managers are scrambling to restore a system of values in the workplace." Is this concern sincere or are corporate executives merely paying lip service to the notion of business ethics? And does it include the ethical responsibility to conserve natural resources and protect environmental values in the pursuit of the free enterprise system? Only time will tell, but as Mark Pastin, an Arizona State University professor, noted, "There is no question that all sorts of people are feasting at the trough called ethics."

Third, we must recognize that, while economic deficits may grab today's headlines, environmental deficits will dominate our future. Any good accounting system can gauge the relationship between assets and liabilities. But what accounting system can tell when our environmental deficit is growing beyond our means to reverse the effects of overspending?

Specifically, we are spending our biological capital at breakneck speed without adequately reinvesting in our natural resources portfolio. That overspending can no longer be tolerated. Yet, it cannot be halted permanently until we develop a comprehensive inventory of our global resources and a far more effective foresight capability to help guide society through the process of multiple-option decision making. That will require a great leap in our scientific learning curve.

Our learning curve is so advanced that we can measure the distance between Earth and the moon, which is almost 250,000 miles, and not be off by half an inch. Yet, we do not know how many species of life share this planet with humankind. That must change.

In light of the limited financial resources available and the scale of worldwide destruction of natural resources, it would be prudent to identify systematically the areas of greatest ecological importance and aggressively protect them. Known in medical circles as a triage, this priority-ranking approach would be controversial. But it would force governments, individuals, business leaders, and others to recognize at last the value of natural resource systems, the threats to their survival, and the possible means of protecting them.

As ecologist Norman Myers noted recently, far from seeking to establish quantification of all critical parameters, a triage approach would identify all relevant sets of values in order to illuminate an unduly confused situation. Such an approach would bring a degree of order to the current haphazard process and encourage the best use of limited financial and other resources. By emphasizing the protection of entire communities of species or entire ecosystems, we could avoid the moral dilemma inherent in saving a single endangered species.

Not only do we need to approach research more creatively, we need to move new information into the public policy and resource management arenas more effectively. Relevant research must reach the table where decisions are made.

Fourth, the term national security must be redefined throughout the world. National security must mean more than straightforward military might. While every country needs an appropriate national defense program, every country also needs national policies that will assure the personal security of individual citizens. After all, the long-term threats to human survival do not come so much from military incursions as from global activities that degrade the natural resources upon which every economy and all of life are based.

Finally, all elements of society must work cooperatively to find meaningful solutions to today's complex problems. This means a dynamic, purposeful coalition, composed not just of environmental groups, but including business leaders, research and educational institutions, writers, philosophers, and, especially, local leaders.

As philosopher Richard Kostelanetz has noted, "The crucial question confronting us now is not *whether* we can change the world, but *what kind* of world do we want. For nearly everything even slightly credible is becoming possible . . . once we decide what and why it should be."

As environmentalists and citizens of the world, we have a good understanding of what and why it should be. Quite obviously, in the 1990s and beyond, we face ever more complex environmental problems. In the 1960s and '70s, we fought the pollution that we *could* see, and we won many important battles. But the 1980s brought us the technical expertise to detect toxic pollutants we cannot see. In the past few years, we have learned far more about these pollutants. And we have come to realize that they are more pervasive and more harmful than the pollutants that once dominated the environmental concerns of the 1960s and '70s.

Our growing body of scientific knowledge has made it clear that hewing to the course already laid out will not achieve the changes needed to improve a complex and demanding world. Let me outline two ideas whose time has come.

First, the U.S. Environmental Protection Agency (EPA) should be upgraded to a cabinet-level Department of Environmental Protection. Our national goals of sustained economic development, enhanced environmental protection, and greater governmental efficiency dictate no less.

While I agree with those who predict that local control of environmental quality will—and should—increase in the next decade, I also believe that EPA's role will become more important. After all, no individual, no industry, no region of the country is untouched by EPA's regulations and domain. The nation's quality of life is determined more directly by EPA than by any current cabinet-level department. And the issue of toxic pollution—among others—demands a national approach from a department adequately funded and fully equipped to deal with these life-threatening problems. In its current status as an independent agency, EPA is woefully ill-equipped to deal with the environmental issues of the twenty-first century. Quite simply, when policy decisions are made at the highest levels of government, environmental concerns are neither represented at the so-called big table nor aired adequately in the Oval Office. That is unacceptable. It is a message we must convey to the American people and our society's decision makers.

My second recommendation is probably more controversial, but just as necessary. We need to amend the U.S. Constitution to include language guaranteeing environmental quality.

The national reluctance to amend the Constitution is understandable. It is, without question, among the world's greatest governing documents. The Constitution embodies broad concepts that make our form of government unique and benevolent. And for the past 200 years, it has worked admirably well, having been amended only twenty-six times. (Ten of those amendments, of course, are the Bill of Rights).

So why alter the Constitution any further? The freedoms guaranteed in the Constitution cannot be realized fully if conservation of natural resources and protection of the environment are not guaranteed as fundamental rights of all United States citizens.

The delegates to the National Wildlife Federation's annual meeting in 1987 voted unanimously in favor of the following Environmental Quality Amendment:

> The people have a right to clean air, pure water, productive soils and to the conservation of the natural, scenic, historic, recreational, esthetic and economic values of the environment. America's natural resources are the common property of all the people, including generations yet to come. As trustee of these resources, the United States Government shall conserve and maintain them for the benefit of all people.

The Federation recognizes that the road toward enactment of a constitutional amendment will be a long and difficult one. While we are not assured of success, success is, nonetheless, imperative.

While the Constitution is a map to freedom, it is not yet a complete map. The contour of the terrain—the United States—has changed in 200 years. We need to alter the map, just slightly, to insure that we are headed for a future in which the productivity of our natural resources and the health of our environment are fundamental freedoms.

"Freedom," wrote Schiller, "exists only in the land of dreams." America is that land of dreams. By exploring the realms of individual and social freedoms, America is making its greatest contribution to the modern age. We as individuals are obligated to make our contribution as well.

The dreams, the vision, the goals I have outlined here are ambitious. I have raised some tough and unpleasant issues. Unfortunately, they are real and their resolution is not without risks. But the risk of losing Earth's life-support systems demands our most dedicated efforts.

Finally, as we look to the future, the paths we must follow require a new and inspired level of dedicated leadership. We need leaders who can set aside narrow, provincial thinking and adopt the broader goal of a nation secure in both its economic vitality and the protection of its environment. We need leaders who are willing to take risks, but not with the health or our environment or the natural heritage we hold in trust for future generations. We need leaders to educate society and to provide the scientific knowledge for continued advancement. We need leaders from all walks of life who have an inspired vision of a better tomorrow and a sense of stewardship for those yet unborn and the environment on which all life depends.

13

Managing the Future, If We Want To

Peter A. A. Berle

I GRADUATED FROM LAW school in June 1964. When I became a member of the bar, there was no National Environmental Policy Act, no Clean Air Act, no Clean Water Act, no Endangered Species Act, no Superfund legislation (dealing with cleanup of toxic waste), no Toxic Substances Control Act, no Safe Drinking Water Act. In short, the legislative framework under which environmental protection is now managed at the federal level did not exist.

Bringing environmental issues before the courts was difficult, because the rules made it hard for anyone whose person or property was not physically damaged to sue a polluter. The right of organizations to sue on behalf of their members was not widely recognized, and the ability to bring class action suits was limited. Administrative agencies dealing with environmental protection had not been established. The task was performed, if at all, at the state and federal levels by fish and wildlife agencies or health departments. The latter were concerned primarily with potability of tap water, the relationship between new construction and sewage treatment, or the location of septic tanks.

In twenty-five years, we have enacted laws at the federal, state, and local levels that are designed to provide environmental protection. These laws have an impact on the way almost all government agencies do business, on the way publicly financed construction takes place, on the matter in which many products and foodstuffs are produced, on the way energy needs are met, and, in varying degrees, on the way we use land. We have built large bureaucracies and promulgated untold numbers of regulations that range from the wise to the idiotic. We have provided access to the courts for citizens and groups who are empowered to challenge not only what an agency decides, but how it goes about making its decisions.

The result has been to diminish the power of any particular agency, entrepreneur, or institution to impose its value system (whether well motivated or otherwise) on everyone else. A utility is no longer free to determine that marketing power cheaply, while spewing large amounts of sulfur dioxide into the atmosphere, is acceptable. Auto manufacturers can no longer sell fleets of vehicles without reference to the fuel efficiency of their products. The Army Corps of Engineers no longer has a free hand to channel rivers and drain wetlands.

Gradually, too gradually for many of us, we are making the necessary adjustments in the market system to reflect the true cost of what we produce. Disposing of waste by polluting air, water, and land is becoming less possible. We are beginning to recognize that when the environment is used as a free disposal system, the polluter is destroying public resources for private gain.

Increasingly, the presumption that some people have a greater claim on public resources than others is being challenged. The U.S. Forest Service should consider the interest of commercial loggers no more important than those of others using national forests. Similarly, the U.S. Bureau of Land Management should view cattle ranchers as having no greater or lesser claim to rangelands than recreationists or those dependent on undisturbed watersheds. In the past two decades, one judicial decision after another has made it clear that times have changed and that laws protecting the environment will be adhered to.

Have these developments had an impact? I believe so. Barry Commoner is wrong when he argues that environmental regulation has been
ineffective since contaminant levels have decreased only slightly (with the
exception of a few substances like PCBs where absolute bans have been
promulgated). Between Earth Day in 1970 and January 1, 1988, the U.S.
gross national product increased from $1,015.5 billion to $4,660.9 billion.
In real terms, per capita income went from $8,097 to $13,050. In addition,
U.S. population went from 204,830,000 to 244,966,000 and the number of
Americans working increased from 78,413,000 to 114,129,000. However
the growth is measured, it has been substantial. Two conclusions may be
drawn from these statistics. First, this growth, which occurred simultaneously with increased environmental regulation and standard-setting,
refutes the argument that environmental protection induces economic
stagnation. Second, without environmental controls, quality of life and the
state of the environment in the United States would be far worse, if not
intolerable. Scientific research is not required to prove this point. Simply
breathe the air during rush hour in Athens, Greece, look at pesticide levels
in some Mexican vegetables, or drink the fresh water in much of Africa.
Then consider what damage would have occurred in the U.S. if its growth
had continued without reference to environmental degradation.

Are we protecting the environment well enough to prevent significant
and irreversible changes in the biosphere as we enter the twenty-first
century? No. Change will occur no matter what we do, if for no other
reason than if present rates continue world population will double in
forty years.

The challenge then is to influence the nature of change so that a
healthy Earth survives and we can have a significant impact on that outcome. We can do it by defining our national security in terms of fostering
a sound global environment rather than simply achieving an arms advantage. We can do it through economic activity and aid that assists worldwide
efforts to limit population growth. We can do it by insuring that the varying
demands on our resources are considered before those resources are
taken over by special interests.

We need to do a better job of measuring the benefits of environmental
protection. Every plant manager can describe to the penny how much
control technology will cost, but the benefit of a pollution-free environ-

ment is much less quantifiable. Long-term impact on ecosystems cannot always be calculated. Putting a dollar value on the unknown risk of sustaining an undefined damage is impossible. But the lack of mechanisms to quantify does not mean that environmental protection is without value. When the damage becomes quantifiable, it may be too late to reverse or mitigate it. Since long-term costs and benefits frequently are impossible to calculate or compensate, one cannot make environmental policy decisions solely on the basis of currently anticipated cost figures. Everyone who argues that the cost of some environmental strategy is too high must be asked whether benefits of that strategy have been analyzed, whether all costs and all benefits have been considered, and what costs and benefits are also involved that cannot be measured.

Keeping the global habitat livable will require more than improved techniques for measuring and quantifying environmental impact. Our survival depends on increasing environmental awareness and concern to such a degree that our voting habits, consumption patterns, management decisions, economic judgments, and the demands we make on our government and fellow citizens reflect an understanding of the fragile nature of our ecosystems and the importance of sustainable resource use. In short, protecting the environment depends on how and how hard we think about it. To date, we have not succeeded in conveying the urgency of the environmental message.

Particular attention must be paid to the thousands of citizens who serve on local zoning commissions and planning boards and other bodies that determine land use at the township, municipal, and county levels. More than anyone else, these people are etching the face of the America that will exist in the next century. More must be done to help them acquire the ability to act wisely and effectively.

Looking forward, I believe that the future is manageable. The real challenge is to convince enough people that it can be done, and to help them acquire the wisdom to do it.

14

One Step Forward

John N. Cole

AS A JOURNALIST WHO WROTE several editorials in 1968 and 1969 urging members of the Maine legislature to enact laws with enough sinew to protect the state's wetlands, I was pleased to be able to report that both the lawmakers and Governor Kenneth Curtis had done as requested.

Indeed, the precedent-shattering environmental laws, debated and enacted some twenty years ago by Maine's 103rd and 104th legislatures, continue to be the fulcrum of what little leverage the state can apply to deal with pollution and development. During those yeasty sessions, the governor signed the site selection law, which defines limitations on both industrial and residential development; the bill to establish a new Department of Conservation; the adoption of air- and water-quality standards; the wild rivers act; and perhaps the most stringent of the lot, the wetlands protection bill. Maine's first line of environmental defense was remarkable in itself, but its near unanimous appproval by old-line Republicans and a young Democratic governor with an equally liberal cabinet still stands as a benchmark of bipartisan cooperation.

Perhaps it was the ever-present tensions of two-party politics that endowed these measures with their staying power. Each party worked overtime to make certain the bills would be tightly written, carefully crafted, and buttressed by every legal precedent that diligent attorneys on both sides could discover. Neither Democrats nor Republicans wanted to be the group responsible for measures that gave under pressure.

Indeed, most of those laws and ones passed later have been challenged in the courts. The oil conveyance law, which fined owners and tankers that spilled oil in Maine's waters and asked them to pay for cleanup and damages, went all the way to the U.S. Supreme Court for ultimate certification. As the decades pass, it is the resilience of these twenty-year-old laws, enacted by part-time citizen legislators, that emerges as their most remarkable and sustaining virtue. Given the combined force of the Reagan administration's eight years of trying to disown previously enacted environmental controls, and the constant challenges brought by those who consider them barriers to enterprise, the feisty longevity of Maine's conservation laws is one of the most permanent and glowing tributes ever accorded any of the state's legislative sessions.

Every citizen who cares a whit about protecting Maine's environmental integrity, and they are certainly a solid majority, should take a moment each day to thank the men and women of the 103rd and 104th legislatures. In retrospect, it is the excellence of their achievement that has kept, and continues to keep, Maine from joining much of the rest of the nation in what has become a general backslide from the enthusiasms of Earth Day to the satisfactions of business as usual. That regression would have accelerated without the environmental defenses that lawmakers pounded so firmly into the foundations of state government.

Since early in 1983, encouraged by James Watt and other unmistakable signals from the Reagan administration that the Earth Day party was over, the Bath Iron Works (BIW) has been trying to breach the legal walls those Maine lawmakers built. As the state's largest single industrial employer (some 8,500 workers), and as one of the only three yards in the nation still able to bid for a chance to build the U.S. Navy's new *Aegis* cruiser, the century-old shipbuilding yard on the banks of the Kennebec River in Bath is outgrowing its waterfront site.

As the producer of a record-breaking number of destroyers during World War II, BIW has had a fifty-year friendship with the Navy, now the only source of serious shipbuilding orders left in the nation. And, as the winner of some of the first multimillion-dollar *Aegis* cruiser contracts early in the 1980s, BIW has spent much of the decade expanding its territory. In spite of a friendly administration ready to spend budget-boggling amounts on national defense, and in spite of a secretary of the interior apparently determined to ignore every environmental restriction on the books, BIW has yet to get permission to fill in most of the ten-acre Trufant marsh next to its main yard.

But BIW keeps trying. Three different variations of the same marsh acquisitions theme have reached the public hearing stage in the past five years, and surely other versions have been proposed behind boardroom doors. Those that make their public debut are always the same product of the corporation's need for room, but as the years pass, their costumes change. Now, BIW proposals arrive cloaked in trade-offs designed to unhinge the armor of the salt marsh's defenders—trade-offs like the recent promise to donate $100,000 to the Nature Conservancy to be used to purchase more marshland for the Rachel Carson Wildlife Refuge in Wells, Maine.

"The marsh in Wells has nothing to do with the one in Bath," says Leon Ogrodnik, a retired Massachusetts businessman who moved a few miles from the shipyard just as BIW began its acquisition efforts. A fly fisherman whose obsession with the sport has driven him the length of the East Coast in an old station wagon with an aluminum skiff riding its luggage rack, Ogrodnik has been pumping organizational energy into the Maine Wetlands Protection Association (MWPA) since he cofounded it last fall with William Hennessey, a former Maine legislator and lifetime Bath resident. Hennessey voted for the Coastal Wetlands Act when he served in the Maine House. Hennessey's years of experience with Augusta's bureaucracy and Ogrodnik's commitment are a formidable combination.

Although Hennessey leaves little room for compromise, Ogrodnik is the more outspoken of the two. "The bottom line," Ogrodnik argues, "is that BIW wants a foothold on this marsh. Then they'll pave the whole ten acres."

"They'll use this marsh for a few years," says Hennessey. "Then they'll be building some new ship, or maybe they won't be building anything. But the marsh will be gone. It took 10,000 years to create. It was here when the Indians owned the place. But it will be gone forever if BIW gets what they want."

BIW's latest proposal, the fourth since 1983, asks for permission to construct a road along the inner rim of the Trufant Marsh. Just over a quarter-mile long and more than fifty feet wide, the road would be strong enough to support the 220-ton modular units that would be joined to become an *Aegis* cruiser. Because there is no room in the current yard to store the massive units, argue the shipbuilders, they need the road to haul them to a BIW storage area south of the main yard.

"If they get the road, then they'll take the whole marsh," says Ogrodnik. "They have other storage alternatives, but BIW can't understand why they can't have the marsh. The company has gotten everything it wanted for the past fifty years. Now they've been denied, and they won't take no for an answer."

Ogrodnik, Hennessey, and other MWPA members have fought their battle in the press, writing letters to the editor and guest columns about the importance of the salt marsh to fisheries and shellfisheries, natural flood control, and waterfowl life support, but it is their legal offensive that has shredded BIW's plans. Without the Coastal Wetlands Act, the MWPA and its allies, the Maine Audubon Society and the state's Natural Resources Council, would have fought the good fight and lost. That is their consensus as well as BIW's. In spite of Watt, Reagan, a softening of the national environmental resolve, and Maine's new Republican governor who supports BIW's expansion, the Iron Works has not gotten Trufant Marsh after five years of trying. Even Maine's largest industrial employer cannot edge around, climb over, or knock down the legal protection those legislators gave Maine's coastal wetlands two decades ago.

And that protection is stronger now than it was then. It was reinforced in February 1988 by two decisions of the U.S. Supreme Court. The first verifies the right of Massachusetts to ask a wetlands violator to pay the costs of restoring the damaged marsh; the second certifies a state's right to claim the land in the intertidal zone and reinforces Maine's claim to the Trufant Marsh. As MWPA attorney Jeffrey Thaler of Lewiston argued in a recent brief, "Pursuant to the Court's decision, the coastal wetlands BIW seeks to destroy are owned by the State of Maine in public trust. Consequently, BIW does not have right, title or interest and may not be issued a permit to build a road across the public's lands. . . . Where the marsh at issue is a viable, living one, state law precludes its destruction by filling." Jon Lund, Maine's attorney general when the Coastal Wetlands Act was written and enacted, agrees.

One brief, however articulate, may not resolve the issue. If BIW dogs the case to its ultimate level, legal arguments can reach the Supreme Court. That is unlikely, however. Given its public relations anxieties as a weapons manufacturer in a nation increasingly aware of the high cost of defense, BIW will likely opt for one of several storage alternatives that do not impinge on Trufant Marsh.

Two decades after Maine's Coastal Wetlands Act became law, most observers, including those at BIW, agree that the Trufant Marsh will flourish for a good while longer—perhaps, as the law intended, forever. The striped bass that have begun spawning in the Kennebec River, for the first time in more than seventy-five years, will need the marsh and its nutrients to support a native population, as will the river's sturgeon, alewives, eels, herring, spearing, shad, bluefish, salmon, and the scores of other species that once made the Kennebec one of the most fertile protein producers in the world.

It was, after all, his interest in fish that first provoked Leon Ogrodnik to research the weapons that might be used to defend Trufant Marsh. That he picked up the sword bequeathed to Maine by its legislature two decades ago is a significant testimonial to the social reevaluation that Earth Day ignited. The national and international rethinking of those times is a lasting legacy, an inheritance of environmental values held in trust for the rest of time.

The market for those values will fluctuate under the pressures of politics and public policy. The Reagan-Watt influence continues. Gains in the battles against toxic wastes and for a sane recycling policy are marginal.

The ozone layer is becoming depleted, as are the national forests. Acid rain still falls, and land-use policies get lip service, not litigation.

Even the agencies established twenty years ago are essentially in neutral. In 1971, Maine's Department of Environmental Protection processed 263 development applications. A generation later, it is looking at more than 2,400 land-use applications and has been advised it must add at least 55 new staff positions to the 350 if it has plans on keeping pace with increasing demands. Similar shortfalls persist in the agency's enforcement capabilities; known violators of Maine's considerable array of environmental regulations take comfort in the knowledge that the department has all but suspended prosecutions and penalties.

But those penalties are there to be invoked, as they surely will be when the scale of social values is recalibrated. The laws that are in place will not be dislodged. They are there, as Ogrodnik and Hennessey know, to be used whenever the public conscience and will are motivated. As BIW has learned, a minority of two can change the course of their state's largest and most intimidating industry.

Earth Day raised our environmental consciousness. But it was the U.S. Congress, state legislatures, city councils, town boards, and village selectmen who responded with laws, ordinances, regulations, and codes, reaching from Maine to California, each designed to slow the destruction of the resources that sustain the planet we have perceived so recently as fragile.

Like governors and mayors, presidents come and go. Since those heady days in the late 1960s when the nation took two giant environmental steps forward, we have taken a step back. But we are still one giant step ahead, and the footprint is there in our laws. In a relatively short time, we will take another step and regain the lost ground.

15

Environmental Quality as a National Purpose

Huey D. Johnson

THE ARRIVAL OF serious air quality problems and the constant news of toxic substances in our drinking water are putting an end to the century-long debate about whether our nation and globe are threatened by environmental decline. The death of Germany's Black Forest, for instance, demonstrates that nature cannot always heal itself. By the time it was understood widely that this famous forest was ill, it was too late to revive its trees.

At its current pace, the environmental movement at best is staying even with growing threats. To solve the problems requires a serious change of approach—an approach described below. The problem is much larger than the movement is organized to solve, but it is solvable. If the environmental movement can reach a much larger audience, its principles will guide the nation and the world to a long-term future. But time is precious.

217

If this nation is to avoid mass destruction and have a future at all, it must respond now. The environmental movement can lead the way by announcing victory and then launching a campaign to educate the public. I believe this declaration of victory is timely and that we can succeed because we who care are now the majority. From corporate chiefs to school children, the concern is becoming unanimous. By unifying these new environmentalists we can become a permanent force. With adequate funding and universal acceptance this movement can equal health, education, and military spending as a national priority.

This call for mobilization to achieve a force far stronger than what the current environmental movement represents does not suggest that we stop what we are doing as environmentalists. Generally, we are a movement of specialized groups—mostly seekers of natural area and environmental quality. This is all to the good. But until we achieve the level of national priority for the environment, much of the movement will remain trapped in the trenches. Where we need the money to solve millions of acres of soil problems, we will continue to have only a few shovels and lots of hopes.

So it is time not to relax but to work harder, and on a far broader and more complex scale than previously attempted. We need to broaden our shared theme, while still retaining our individual narrow focus. It will take a lot more education, advocacy, and demonstration in all interest areas.

My proposal has advantages for individual environmental groups. Each needs more clout and greater financial backing. To realize the need for an expanded scale of effort, we need only look at our weaknesses. Until now we have been splintered enough on organizational issues and goals to keep from being a unified force with which any administration would have to deal.

It is a real hope that Ronald Reagan will be the last president elected—and James Watt and Donald Hodel the last interior secretaries appointed—with their lack of knowledge or concern about the state of the world's environment. But the reality we have to deal with suggests that, left to drift, we are going to get closer to disaster before we as a nation will react in a serious way to solve environmental problems.

Power will come when we join forces, enlarge the circle of those who understand and are supportive, and expand the vision to deal with the complexity of our environmental problems. In this way, the seedbed will be set for planting a new destiny—that of making the restoration of environmental quality a national priority.

Why is it that we can announce victory? For two reasons: The public is now with us, and economics have reversed and joined our side of the argument.

Winning the minds of the people is the major reason for announcing victory. Our efforts have had an enormous educational impact on the public's perceptions about its environment and its dependency on natural resources.

Most Americans are now ready to support the restoration of environmental quality as one of their highest goals. We know this because environmental quality consistently ranks high when Americans' interests and concerns are polled, because of widespread media attention, and because of the changing attitudes of politicians. We have gone mainstream and beyond.

In-depth pieces appear regularly in the front pages of the dailies, and environmental stories often lead television and radio programs.

What could be considered a true litmus test is that the issues have invaded other so-called special interest arenas. For example, two of the best articles of 1987, on the most complex issues we face—water and air quality—appeared in *Business Week* ("Troubled Waters," October 12) and *Sports Illustrated* ("Forecast for Disaster" by Robert H. Boyle, November 16).

There are clear indicators that the message of limits has hit home. Suddenly, within recent months, most politicians have become environmental advocates, in part because the winning arguments are on the side of environmental quality.

The outcome of the last energy crisis is a prime reason. When conservation was proposed as a solution to immediate and long-term energy problems, the traditional sellers of energy—utilities and oil companies—tried their best to argue otherwise. But the quick success of conservation in the face of their intense opposition was instrumental in winning public support. And more significant is that many of those utilities and oil companies have since joined us—another way of measuring our readiness to act on a higher and larger scale.

For example, in the April 1988 issue of *Scientific American,* northern California's Pacific Gas & Electric Company admits that "Conservation will allow us to avoid $5 to $7 billion of outlays for new capacities that would otherwise be needed in the next decade. It costs up to seven times as much to produce one kilowatt hour from a new energy source as it does to save one kilowatt hour from PG&E's conservation programs."

Another good example of what is made possible by including economic factors, while not letting economics dominate the environmental philosophy, occurred in 1978 while I was serving as secretary for natural resources in California. We developed one of the first comprehensive programs for resource renewal. By encouraging diverse interests, including industry, banks, agriculture, labor, and environment, to join forces and to sign on to the program, the resource agency was able to get $400 million over four years from the legislature to carry out the Investing for Prosperity program. Many of the thirty-five resource renewal projects are still being funded due to their success.

One of the projects assisted private timberland owners in California. The state had 13 million acres of timberland, but only 8 million were forested. The other 5 million acres needed to be planted and managed, but these owners of up to 5,000 acres could not afford to go to the bank and borrow the money needed to do so.

Five million dollars were invested in these timberlands in the first six years of the program. The money came from publicly owned resources—principally timber revenues from state-owned land.

According to a study requested by the current administration in California, the results, projected over fifty to seventy-five years, found that the one expenditure of $5 million will return more than $400 million in new timber sales and $100 million in new taxes and provide 18,000 additional jobs.

These figures were the result of one economist's study. Others' figures may vary, but the value of the forestry project is indisputable.

A qualifier should be added here. It is still true that many traditional economists do not as yet consider environmental or natural resource factors in their economic forecasts. But many of the younger economists coming up do, and they represent the future.

Broad public awareness is the reward for the hard work of the many environmental groups. The Sierra Club, National Wildlife Federation, Defenders of Wildlife, Wilderness Society, Trout Unlimited, Audubon Society, National Geographic Society, and Cousteau Society, to name just a few, through activities, direct mail, books, magazines, and films, have done much to encourage people to develop an appreciation of nature and then the desire to defend it.

Other groups, such as Greenpeace, Earth First!, and Friends of the River, have concentrated on creative conflict to impress on the public a sense of quality and permanence in landscape and wilderness preservation. And the Natural Resources Defense Council and the Environmental Defense Fund have fought to strengthen legislation and enforce regulatory action. In all these different roles, the movement has been largely victorious.

A hundred thousand successes have brought us to this point of announcing victory. From the hours teachers and volunteers have spent with children on nature walks; to the struggles by community groups for open space and against development that would overwhelm the capacity of their communities; to the eloquent backing of writers, artists, and film makers; to the fortitude of the specialized environmental groups that carried the major policy battles from Redwood National Park to the National Environmental Protection Act, every effort is precious to the overall picture.

Perhaps the most important element in analyzing success is the volunteer hours put in by little-known idealists who kept things going. I have witnessed dying cancer patients using their last energy to appeal to others for help on environmental issues. I have seen staffers in ragtag new environmental groups forgo pay without complaint, because of the importance of the cause.

Where the movement was outspent on any struggle, volunteer persistence made up for the lack of money. The opposition, on the other hand, had to hire employees to do their bidding. The imbalance was in our favor, because no single exploiter could match the volunteer hours and devotion to bringing our message to the public.

Possibly the most effective educational endeavor is the simple process of training. In a number of states, school children, often sixth graders, are required to spend at least a week in outdoor education. The practice has been going on long enough that in several states, including California, the children who participated in those first environmental classes are now voters whose children go to camp.

Having announced victory and having won the struggle for the majority's awareness—with corporate and political leaders now considering themselves as environmentalists—what do we do next to mobilize? I believe we need to select a universal theme that can encompass past ideas as well as ideas and solutions yet to come. I see the restoration of environmental quality as a national purpose, one that will be accomplished by investing in a nationwide hundred-year plan of resource recovery.

To begin, a team working out of a national organization would assist each state in defining its own hundred-year plan. In time, as restoration quickly becomes a national purpose, these plans and subsequent programs would require the investment of some of our national capital in delivering our own future.

This comprehensive, in-depth plan, which I call the Green Century Project, would have as its ultimate objective a healthy environment and preservation of natural resources to pass on to future generations. We should keep in mind that the second generation of the environmental movement needs to be involved in developing an overall plan that enlarges the scale of vision, formally announces victory, and defines our preferred destiny in a way that the public can understand and support.

Of major importance will be determining where the money will come from to restore environmental quality, in order for natural resources and the environment to rate with roads, education, health, and military spending. At present only 3 percent of federal taxpayers' money goes to environmental and natural resource issues.

Meanwhile, half our tax dollars go to defense-related expenditures, with no return on our investment except an imagined wall of security. A key issue on which we must take a stand is that national security is in large part internal. History shows us that a hungry or threatened populace rebels against whomever is in power. We need to get the real message across to the people, that environmental quality and healthy resources are national security issues. Greening the land with new forests will improve air, cut back on carbon dioxide, conserve energy, add beauty, and return jobs and much needed resources to a flagging economy. Soldiers can plant forests if need be.

The new reality includes the links between resources, between projects and policies, and between the causes and effects of human activity. Our movement will not be fully effective until it takes on the obviously linked issues it tries to avoid. An example is population and carrying capacity, a root cause of our global problems. Carrying capacity is the ability of a region to sustain a community of living organisms, of which people are one species. In a large state like California, for example, whose population increases by half a million-plus every year, its southern residents have run out of air and its northern residents are low on water. And statewide there is traffic congestion and limited affordable housing. The solution to these problems will require policy changes in resource allocation that are considerate of carrying capacity and the public trust.

There are presently few foundation sources for the money it will take to launch the mobilization effort to restore environmental quality and achieve the power we need to make it a national purpose. This is because foundation policy avoids conflict, and conflict is an inherent process of resource allocation. There is a strong possibility that foundation policies will change to support this new national purpose direction as more individuals get involved and make a commitment to environmental quality, in all its complexity.

We need to focus primarily on the complexity of issues in this country if we are to deliver a more positive environmental future anywhere. And we need to push the politicians to be involved and establish a policy whereby a share of national assets will be invested in recovering resource wealth. That will only come when the voters vote for a leader willing to take the risk, the leap of faith needed to grasp the opportunity to move the nation into permanence.

Unfortunately, there has not been a political leader with this needed vision since the beginning of this century, when Theodore Roosevelt appealed for natural resource renewal and environmental quality as a national priority. But that is another reason for a broadened involvement in the environmental movement. It is a political rule that if a parade forms, a national leader will emerge to lead it.

We must begin the parade.

16

A View from the Trenches

Cynthia Wilson

IN 1965, WHEN I MOVED to Washington, D.C., the Potomac River was an open sewer. On summer days, driving across Memorial Bridge in the area where the John F. Kennedy Center for the Performing Arts now sits, one was assailed by the stench. Anyone who accidentally fell into the river was warned to get a tetanus shot.

Today, thanks to a massive infusion of federal money under the Clean Water Act to build sewage treatment plants, the Potomac no longer stinks and adventurous residents wind surf in the river, while some brave souls even eat the fish they catch!

However, the nearby Chesapeake Bay—the nation's most productive estuary—is dying, the victim of chemicals and land development in Maryland, Virginia, and Pennsylvania. While an ambitious joint state-federal effort is underway to try and save the bay through a variety of land use controls, the outcome is uncertain.

A mix of victories can be pointed to in other issues as well. The effort to save the endangered alligator has been so successful that the animal is now considered a nuisance in some places. But the last wild California condor has been taken into captivity and the future of that species is in doubt.

Pittsburgh's air is cleaner, mostly due to the closure of obsolete and inefficient steel plants. But the ozone layer is thinning at a dangerous rate, and unless we stop using chlorofluorocarbons, some scientists believe that each 1 percent decrease in ozone layers could cause an additional 20,000 cases of skin cancer annually among Americans alone.

The scorecard would be far worse if the environmental movement had not been active, but the nature of the challenge has not changed. The National Environmental Policy Act and a series of new laws gave us tools to block development, force the cleanup of certain pollutants, and otherwise slow down the desecration of the earth. But those laws are not all working, and as we bid farewell to the Reagan era we should reassess our goals and tactics.

In the late 1960s, most national conservation organizations had only a few people who actually worked on legislation and dealt with federal agencies, primarily the Interior Department. Many of the long-time staffers had been trained as wildlife biologists or foresters. Others had a journalism or some other type of liberal arts background. When I was hired by the National Audubon Society in 1969 to open a one-person Washington office, the only other woman "professional" working in this milieu was Lois Sharpe, the long-time water expert of the League of Women Voters.

In part because of fear of losing their organizations' tax exemptions for lobbying (as the Sierra Club had) the groups kept a low profile on legislation and were generally unsophisticated politically. For the most part, the grassroots members of these groups were not organized for political action, and like most Americans did not understand how Congress really works.

The growth of the movement following Earth Day brought an infusion of younger people with liberal arts, political science, or law backgrounds, and as various universities developed more diverse natural resources programs, their graduates migrated to staff the organizations, which now began calling themselves environmental. In retrospect, it is amazing how

226

much important legislation got passed in the early 1970s by this relative handful of paid staff and a growing, though still relatively unsophisticated, grassroots effort. I think the answer is that we were still enjoying the honeymoon period, and the chickens had not yet come home to roost.

By contrast, with today's large, skilled Washington staffs and well organized grass roots memberships, the environmental movement has not been able to break the stalemate on issues such as acid rain, in part because of the hostility of the Reagan administration as well as complex issues such as the impact of tougher air pollution regulations on the coal miners of West Virginia. Some critics have concluded that the movement's focus is the problem and that the national organizations have become too much like the establishment.

I believe that view is naive. The truth is that the days of easy victories are over. Once the impact of regulating pollution was felt, the backlash set in and industry learned how to block the environmental community's efforts effectively. Today, we need an effective presence at the federal, state, and local levels to deal with the immense problems we face. There is more than enough work for everyone to do.

During 1986–87, a spate of articles in publications ranging from *Mother Jones* to *Esquire* focused on differences in tactics between the direct action groups such as Earth First! and the more mainstream groups. (I found it amusing that Friends of the Earth—my organization—was considered in the latter category.) Some writers, such as Barry Commoner, were highly critical of the environmental movement, and a fair amount of energy was spent in rebutting some of the accusations that appeared.

While discussion of goals and tactics is healthy, I believe that if we are to succeed in preserving a habitable earth, we must focus our energy on the real enemy and quit pointing fingers at each other. There have always been various opinions on how best to fight environmental battles. Although Friends of the Earth (FOE) is committed by its bylaws to lawful methods of activism, we do not think anything is gained by attacking others who use different tactics.

Different approaches lend themselves to different problems. Spiking a tree or sailing in front of a whaler gets media coverage, which may help to focus the attention of the public and responsible officials on the problem at least temporarily. But that is only the first step, and other, less visible work is usually required to solve the problem. Long years of work with the International Whaling Commission and building public opinion have been needed to bring significant reductions in whaling. Lobbying Congress and filing lawsuits against the U.S. Forest Service have saved more trees than "ecotage."

If we view the cause as an ecosystem in which everything is related, few issues are truly local. The toxic chemicals that are dumped in New Jersey will find their way into the environment in one form or another and add to the total toxic burden. The wandering garbage barge from Islip, New York, was a messenger, telling us that it is time to face up to the issue of solid waste.

In the 1970s, the focus of much of the work of the environmental community was on the national level. It was also a period of tremendous growth in membership. To varying degrees, the national organizations worked to train their members in the legislative process, and, for example, it was those grassroots members writing and visiting their congressmen who ultimately made victories such as the Alaska National Interest Lands Conservation Act possible. (The active support of the Carter administration was also a major factor.)

During the dismal Reagan years, the resources of the national groups have been concentrated on preventing the erosion of major laws and the action has shifted in part to the state level. For example, the impetus for passage of landmark legislation to try and save the Chesapeake Bay came from the state of Maryland, not from the federal level.

In a number of cases local governments are taking novel though sometimes mainly symbolic steps. For example, Takoma Park, Maryland, has declared itself a nuclear-free zone while Berkeley and a number of other California communities have banned Styrofoam. Such efforts are helping to keep these issues before the public.

An example of a regional effort is the Northwest Power Planning Council's Protected Areas Program. Using a process established by federal law in response to the power surplus in the Pacific Northwest, FOE has served as a catalyst for grassroots efforts to press the council to protect thousands of miles of rivers from new hydroelectric dams. By working directly with the council staff as it amassed a four-state inventory of river resources, and providing information about the process to a network of interested individuals and organizations, FOE's Seattle, Washington, staff has built a groundswell of support for bold action by the council. Despite opposition from private utilities, an impressive coalition of sportsmen, recreationists, Indian tribes, and even the Bonneville Power Administration is supporting the protected areas program. The outlook for success is good. Instead of the endless dam-by-dam battles, which are so draining on resources, this effort will protect the rivers in the whole region.

Environmentalists have been more successful at passing laws than getting them implemented as illustrated by the 1977 strip mining legislation. In part, this is because the legislative battles are, in many ways, more fun to work on, at least as long as your feet hold out! Legislation lends itself to grassroots pressure. The regulatory process that follows is tedious by comparison, lacking in glamor, and ultimately dominated by specialists who talk in jargon about section numbers of laws, such as 404, which is unintelligible to the general public. These long, drawn-out efforts to implement new laws are hard to sustain, yet very important.

During the Reagan administration we have seen a deliberate effort to weaken the laws we have (although admittedly some of them are defective and deserve amendment) and a failure to enforce them. Just when the impact of the new pollution control statutes was being felt, the economy was worsening—creating a jobs-vs.-environment issue that management gladly exploited.

While Anne Gorsuch Burford, former head of the Environmental Protection Agency, and James Watt, Reagan's first Interior Secretary, were stymied in some of their most outrageous efforts to rewrite the laws, they succeeded in achieving their goals by starving out enforcement programs and hiring people such as Rita Lavelle who were eager to accommodate their friends in industry. It is ironic that an administration that has prided itself as being for law and order has not enforced laws more effectively. The next president should appoint a tough prosecutor to head EPA—someone who will throw the book at polluters, and if that does not work, throw them in jail.

Perhaps the longest range damage inflicted by Watt and Burford was driving out of government civil servants who cared about the environment and wanted to enforce the laws. I am not talking about political appointees, but about the career professionals who have been silenced or seen their work rendered ineffective. In the Interior Department, career professionals who had worked on the Alaska lands legislation were some of the first targets of Bill Horn, former chief aide to Alaska Congressman Don Young, who became assistant secretary for fish, wildlife, and parks. While civil service regulations made it difficult to fire these people directly, most were transferred to undesirable jobs ill-suited to their skills or personal needs. Many eventually left government. Those who remained were demoralized and unable to function effectively. Even secretaries who had been with the department since the Nixon administration were shuffled off to less desirable assignments. This damage will be very difficult to repair, yet little attention has been paid to it.

Partly because of EPA's slack enforcement efforts, the environmental community has attempted to bridge the gap. FOE and two other groups brought thirty-eight suits under the citizen suit provision of the Clean Water Act against a number of industrial polluters in New York and New Jersey and won fines that on the average were four times higher than EPA and the Justice Department had been settling for in comparable cases.

Although we have changed radically the way government and the public look at the environment and slowed the pace of destruction, it is clear that there has been irreversible damage. In many cases, what is needed is simply to continue doing what we have been doing—holding the line to prevent congressional backsliding, tedious monitoring of existing law, and continuing to maintain pressure on government and the corporate sector to clean up their act. But some statutes, such as the pesticide law (FIFRA), are so deficient they need a complete overhaul—which thus far has eluded the community.

In 1989, the newly elected president and Congress have an opportunity to renew the quest for a healthy planet and act boldly to correct those problems that are within our power to correct. Nuclear war is the ultimate environmental threat, but even without it, the prognosis for the biosphere is not bright, unless major changes are made.

We have made irreversible changes already—most notably in the loss of species and the alteration of the landscape. To date, we have been miserable failures at restoring the damage we inflict. During the years of fighting to enact the 1977 strip mining law, one of the major issues was a requirement to restore the lands. Yet more than 3,000 surface coal mines have been left unrestored, and the federal government has failed to collect more than $22 million in penalties, fines, and fees for violation of the law.

Perhaps the technicians will develop technologies to patch up *some* of the damage we do, but we cannot bring back the passenger pigeon. Even if the monkeywrench gang blows up Glen Canyon Dam and drains the lake behind it, the canyon would not be the same as some remember it.

More frightening is the climatic change resulting from the greenhouse effect and the impact on health from depletion of the ozone layer. If the scientists are correct, those climatic changes are already underway and we cannot reverse them. It is possible that a herculean effort may slow them down, but once put in motion the forces of nature are still beyond our control.

We do have an opportunity to stop destroying the ozone layer through banning chlorofluorocarbons (CFCs), but whether the nations of the world have the will to do so remains to be seen. By itself, the treaty signed by thirty-one nations in 1987 only reduces the use of CFCs and will not do the whole job. But it is a first step.

On the brighter side, in the past twenty years the environmental movement has changed the way we relate to nature and view the world. People are no longer incredulous over efforts to save an endangered butterfly, although saving warm, furry creatures is still likely to garner broader support. Still, today many people—especially those who have grown up since Earth Day—recognize that the butterfly is a symbol of the global need to save biological diversity. They understand that in saving the butterfly, we are also saving its habitat and other creatures as well.

A number of years ago, a colleague of mine at the National Audubon Society commented that he thought the California condor was not going to survive, because its numbers had already fallen so low and it had such difficulty coexisting with man. But he felt all our efforts to protect it and to set aside its habitat were just as important, given the population development pressures in California.

The arguments continue over the value of capturing the last wild condor in the hope of breeding the species in captivity, but at least people are concerned about an ugly vulture—which says that we have come a long way. If we can learn to love this creature, perhaps we can become

231

more appreciative and tolerant of diversity in other species—including man.

What I find most encouraging is that more and more people are viewing issues in the global context and can readily see why it is vital that we save tropical rain forests—not just for the fuzzy monkeys and vivid birds, but for the air we breathe. Educating the consumer to make the link between a teak dining room table and the destruction of rain forests is our next challenge.

Helping indigenous people in Third World countries resist their governments' efforts to import nuclear technology may help them prevent another Chernobyl. Friends of the Earth is putting more and more of its energy into joint campaigns with its thirty-three affiliates around the world, because most environmental problems must be viewed in a global context. But at the same time, we are working here at home to take on the challenges in our own backyard.

Even though the record of the past two decades is mixed, we should not be discouraged. Having been in the trenches since the mid-sixties, I am proud of what the movement has accomplished but realistic about its frailties. Rather than focusing on our failures, now is the time to renew our commitment and go forward, taking advantage of the lessons we have learned along the way. The environmental movement has made an enormous difference, but it cannot stand still and we must avoid the internecine warfare that has crippled many good causes. We must also resist the temptation to be self-righteous. Even though I believe with all my heart and mind that we are on the right side, we need to learn to laugh at our foibles and get back to work.

17

Looking Backwards

Nathaniel P. Reed and Amos S. Eno

AS WE ENTER THE TWILIGHT MONTHS of the Reagan administration and the ebbing years of the twentieth century, it is appropriate to reflect on the progress of the environmental movement and on its failures. The first thought that comes to mind is that current environmental leaders are overly focused on a self-serving agenda, and the lessons of both the immediate past and recent history are relegated to a deep, dark past of irrelevance. Bill Moyers has observed: "Americans have never lingered long looking backward. We're a people of the future. The horizon compels our gaze, not landmarks littering the past. As my mother often said to me, 'Be sure your headlights are brighter than your taillights.' But something is always bumping me from behind, trying to get my attention."

In recent years the environmental movement has been characterized by a pervading climate of frustration. People decry the lack of progress and leadership and speak disparagingly of the federal government's administration of environmental programs. Much of this frustration stems from the notion that progress is continual. How easily we forget! The following paragraph is taken from Stephen Fox's recent biography of John Muir and the history of the conservation movement.

> . . . ran on a platform that called for "restoration of the traditional Republican lands policy," meaning free enterprise without Federal meddling. Once in office, the . . . Administration in its first term compiled a conservation record so execrable that it reinforced the recently forged alliance between conservation and the Democrats. Top positions in conservation bureaus were taken off Civil Service and given to dubious political hacks.
>
> After two years under . . ., it is clear that the Administration is the aggressor, and the Congress the defender. This is not a political opinion, it is a fact.
>
> Conservationists were especially appalled by Secretary of the Interior. . . . For his friendliness to private interests he was called. . . .

The foregoing chronicles the malfeasance of a Republican administration. It is not President Reagan and Interior Secretary James Watt as one might suspect, but rather President Eisenhower and his egregious Secretary, Douglas McKay, the infamous "Giveaway McKay" of the 1950s.

When one looks back over the history of the conservation movement, it is clear that our progress has been two steps forward and one step back. Episodes of progress and initiative characterized by feverish legislative activity and institution building are followed by periods of retrenchment and reaction. The political progenitors of modern conservation, Theodore Roosevelt and his chief forester, Gifford Pinchot, were followed by the conservative President Taft and the scandalous Interior Secretary Richard Ballinger. The New Deal brought renewed energy and innovation to conservation programs under the tutelage of Franklin Roosevelt and Secretary Harold Ickes and their lieutenants Arno Cammerer and Horace Albright for parks and Ding Darling for fish and wildlife. This progressive era was followed by the retrenchment policies of President Eisenhower and his feckless Interior Secretary McKay. The modern era of environmental activism emerged from the dark ages of the '50s (which anticipated a similar attitude in the 1980s) with the Kennedy/Johnson administration under the tutelage of Secretary Stewart Udall.

The momentum built during these democratic administrations was carried forward into President Nixon's tenure. The enactment of a variety of environmental laws coincided with the environmental movement's taking center stage in the arena of national affairs. Most national conservation organizations expanded in size phenomenally and became truly national in scope. But the movement's wave of popularity crested, and by the mid-1970s its influence ebbed and then receded under the complacent tenure of President Ford. Even his environmentally sensitive successor President Carter and his superb Interior Secretary Cecil Andrus could not keep the attention of Congress and the public focused on the need for environmental reform.

Part of the problem stemmed from the increased size and professionalism of environmental lobbying staffs. As *Washington Post* writer T. R. Reid noted: "Public interest lobbyists . . . by becoming commonplace, have lost their cachet. Today, no matter how noble the organization they represent, they are just lobbyists. They take their place in line with everyone else." What Reid overlooked is that "people learn from defeat more than they learn from victory," as Freeman Dyson has wryly commented. The environmental movement rode the wave of success throughout the 1970s without understanding many of the underlying social trends then evolving. President Reagan was more astute in calling for a shift in legislative decisions from Washington to state and local governments.

There are two points that stand out in this historical overview. The first is that the recent preoccupation with national environmental agendas sees the history of the environmental movement in a linear fashion from the so-called dark ages of the nineteenth century to an anticipated age of enlightenment at the end of the twentieth century. Not only is this assessment inaccurate, it ignores the real progress made during periods of retrenchment and obscures the movement's failures during periods of progress. Second, the inconsistencies in environmental policy over time, and particularly during the Reagan administration, have caused disquiet in the national movement. In the past decade, momentum has shifted as the national agenda has dissipated to a more regional, state, and local focus, and this is where the action is today. Unfortunately, a majority of the best-known environmental groups that fought the good fight in the '60s and '70s have become preoccupied with and rigidly focused on their own agendas, which are often remote from local issues.

The environmental movement's preoccupation with the present and immediate future has created a situation where hard-learned lessons are often forgotten and real successes go unacknowledged. This tunnel vision is not limited to the environmental movement, as Bill Moyers comments:

> The lack of historical continuity and communication between the generations is, to me, one of the most disquieting features of our time. What is happening today, this very minute, seems to be our sole criterion for judgment and action. And all of our yesterdays have little relevance.

Current fashion among environmental groups dictates dismissing the Reagan years as a wasteland for the environmental agenda. This is both substantively wrong and a disservice to the thousands of professionals who toil anonymously in the federal and state environmental and natural resource agencies. Looking back over the Reagan years, perhaps the single most important factor that comes to mind is the resiliency of our federal agencies. The National Park Service, Fish and Wildlife Service, Forest Service, and even the Environmental Protection Agency and the Bureau of Land Management to a large degree withstood the Reagan onslaught against the environment. The battlements are still manned and the day-to-day job of taking care of business still gets done.

The furor generated over James Watt and Ann Gorsuch Burford and their many adherents also obscures several very fine political appointments that have been overlooked in all the turbulence of the Reagan years. Ann McLaughlin served valiantly as undersecretary of interior in the wake of Jim Watt and labored diligently to bring a degree of balance and professionalism back to the department. McLaughlin took the initiative to resolve successfully that most rancorous of all water development projects, the infamous Garrison Diversion. Bill Ruckelshaus similarly restored a sense of equanimity, respect, and programmatic initiative in EPA after the Burford episode. Perhaps the most unrecognized environmentalist in recent years is Peter C. Myers, deputy secretary of the Department of Agriculture, who was one of the principal architects of the 1985 Food Security Act, better known as the Farm Bill.

The Farm Bill may be the single most important land conservation law passed during this century. This legislation has not received the attention from the environmental community that it deserves, in large part because it falls outside the focus of the agendas of the majority of national conservation organizations. However, it is hard to overstate the bill's value as a precedent and its long-term implications. In 1985, for the first time in half a century, a Conservation Title (Title XII of the Food Security Act) was woven into the fabric of federal farm support programs to stem excessive soil erosion, enhance water quality, and restore and improve fish and wildlife habitat. The Conservation Title of the 1985 Farm Bill is the strongest piece of soil conservation legislation in U.S. history and the strongest environmental statement ever included in agricultural legislation. Most important, when our country has almost 400 million acres of public lands under federally protected status, the Farm Bill targets private lands for conservation. It does for private lands what the Land and Water Fund did for public lands in 1964. The four principal provisions of the Farm Bill—the Conservation Resource Program, through which 27 million acres of fragile land have already been retired from production; the Sodbuster and Swampbuster provisions; and the conservation compliance program—are all being implemented effectively at the local level.

This brings us back to what we perceive as the biggest failing of the environmental movement—its inability to respond to issues at the local and regional levels. The movement has been preoccupied, with its national focus, with Washington initiatives and with building a system of public natural resource reserves, and in the process the folks back home have been neglected. Too often in recent years the leaders of the major national environmental groups have forged and pursued their own personal agenda without taking into consideration what is happening back home on the farm. Until the 1985 Farm Bill, the most fertile fields for innovative conservation, the private sector, have been ignored. In 1987, 1000 Friends of Florida was incorporated to address the major growth management and natural resource protection issues in Florida, because no one else was paying them any attention. The political powers necessary to address controversial environmental issues are most effective, as every politician knows, if they emanate from the grassroots. Today the agendas of national environmental groups increasingly ignore local concerns.

Nothing symbolizes the failure of the environmental community more starkly than conservationists' inability to build a viable, nationwide political constituency to support annual appropriations for land acquisition under the Land and Water Conservation Fund (LWCF). For too many years the fund was taken for granted and environmental groups simply assumed that Congress and the administration would provide adequate funds to sustain park, refuge, and forest acquisitions. Environmentalists' preoccupation with focusing on Washington-based institutions, instead of building local infrastructures and constituencies to support the fund, enabled the Reagan administration to abolish the fund's coordinating institution, the Heritage Conservation and Recreation Service (formerly the Bureau of Outdoor Recreation) and to terminate the state grant portion of the fund.

A analogous situation has occurred over the past thirty years with respect to our waterfowl populations. Today, the populations of most of our best known waterfowl species, the mallard, pintail, and blackduck, are at their lowest levels in history. This is in spite of more than fifty years of energetic conservation activity on behalf of waterfowl. More dollars and effort have been applied to the conservation of waterfowl of North America than any species of wildlife worldwide, and yet today these species are in desperate straits.

Here again the problem has been one of focus; in this case, environmentalists assumed that the U.S. Fish and Wildlife Service and its state counterparts were doing their jobs. They were, but their purview was limited and on private lands, particularly agricultural lands around the United States and Canada, wetlands were converted at fantastic rates throughout the 1970s and 1980s, sweeping waterfowl and other wetland-dependent species' breeding and wintering habitats off the map.

A third final example of the shortcomings of national environmental groups' recent focus can be seen in their treatment of the President's

Commission on Americans Outdoors, which represents yet another failure in political strategy based on an insular, Washington-based vision. Rather than perceiving the commission as an opportunity to set the agenda for conservation for the next two decades, as the commission's predecessor, the Outdoor Recreation Resources Review Commission, did so successfully two decades earlier, the national environmental community largely ignored the commission because it was set up under the auspices of President Reagan. No matter that the chairman of the commission, Governor Lamar Alexander of Tennessee, is a noted conservationist and that other key conservationists, including Pat Noonan of the Conservation Fund and Gil Grosvenor of National Geographic, were driving forces behind the commission's work. The commission's final report was better than any agenda put forward by the "Gang of Ten" (as the major national environmental groups are called) because it was soundly rooted in the recommendations of local organizations from across the country. The report farsightedly and purposely emphasized the involvement of state and local activists as part of a strategy to build the ever-important political constituency necessary to sustain environmental programs on a nationwide basis.

National environmental groups today are still fighting the last war. They are preaching an Armageddon of ecological disaster around the next corner. Their focus is singularly on Washington and the fixed lines of trenches dug by the major national environmental organizations battling the Reagan administration. Today, the real battle lines are no longer in Washington. The battles are being fought on myriad state and local fronts as in my own state of Florida and Dade County, where the national environmental groups are increasingly irrelevant to the political and social solutions being formulated to contend with resource conservation issues.

We are not overly pessimistic about this state of affairs because the progress and initiatives being developed at local levels across the country more than buoy the misdirected energies of many national environmental organizations. As we move forward toward the next century and plot our strategies for the future, it is imperative that we bear in mind the successes and failures of the past. Our current focus is often too narrow. Today, more than at any time in history, we need to broaden our conservation constituency. We need to reach beyond the traditional confines of environmental conservation to embrace all public and private sectors, for the isolation of environmentalists has been their undoing in the past. The environmental agenda needs to be crafted close to home, not in grandiose documents emanating from Washington, and it should incorporate the interests of the whole spectrum of local constituents and business interests.

18

On Grassroots Environmentalism

*Lois Marie Gibbs
and Karen J. Stults*

THESE PAST SEVERAL YEARS have been special for the environmental movement in general and for the grassroots toxic movement in particular. For the first time in years—perhaps the first time ever—there is a public consensus supporting efforts to protect and improve the environment. According to a Louis Harris poll, this is how the American public defines the most important environmental issues: 92 percent of Americans believe hazardous waste is a serious problem (ranking highest among environmental issues), but also show strong concern over polluted lakes and rivers (90 percent), contaminated water (86 percent), radioactive waste (79 percent), acid rain (79 percent), and air pollution (75 percent). Given my own history at Love Canal and the work Citizen's Clearinghouse for Hazardous Wastes (CCHW) does, I am pleased to see these poll results, because they reflect my ranking of environmental priorities.

These numbers show a new consensus for the environment. They reflect a big change from public attitudes in the mid- to late seventies when environmentalists were mocked as being so-called save-the-whales types, as more interested in protecting the snail darter than caring about human beings, or as hippies or communists, as antinuclear power demonstrators often were characterized. We do not often hear anymore the sentiment stated in the popular '70s bumpersticker: "Hungry? Out of Work? Eat an Environmentalist!"

The main reason for this turnaround, as we see it, is that in the 1980s more attention has been devoted to building a grassroots base of popular support for environmental causes. This is a reverse of the pattern of the 1970s when environmentalists treated the public with benign neglect at best and, at worst, with arrogance. When environmentalism is recast as *environmental justice,* it is easier to see the importance of paying attention to public perception. That is because we can then draw on the history of other movements for social justice and see change and progress as the direct result of winning the battle for the hearts and minds of society.

Two events changed the nature of environmentalism in the 1980s and are largely responsible for the surge in public support. The first was Love Canal (1978), which showed people that toxics (and similar environmental hazards) could turn up in anyone's back yard. The grassroots movement (1981) that followed the uncovering of Love Canal was populist and would not allow the issue to be compromised, bureaucratized, or intellectualized out of existence. The other catalyst was Ronald Reagan's election. The Reagan administration was so extreme and so outrageous in its approach to environmental concerns that it made true believers of even the most disinterested bystanders.

We still face many threats. The United States still produces more than a ton of toxic waste for every man, woman, and child. This does not even include the tons of waste produced by the U.S. military, the single largest toxic generator, with an annual waste output higher than the five largest chemical companies combined. The U.S. General Accounting Office estimates that there are between 130,340 and 425,380 suspect toxic sites (GAO report #RCED-88-44, Dec. 1987). The infamous garbage barge that sailed up and down the Atlantic Coast in search of a dumping ground brought international attention to the problem of municipal solid waste. These specific examples reflect the general attitude of industry and government toward other environmental issues, such as acid rain, protection of endangered species, and preservation of vital natural resources.

As the public becomes aware of these threats, it will begin to work on behalf of positive solutions. When CCHW was founded in 1981, its main goal was to end land disposal of hazardous wastes by the end of the decade. It looks as if this goal will be achieved. We also have fought for positive alternatives—the "Four Rs" of recycling, reduction, reuse, and reclamation. Our colleagues told us that this was a pipe dream, because (a) it's got to go somewhere, and (b) Americans are lazy, stupid, and easily swayed by the power of corporate and government institutions with a vested interest in the status quo.

Yet, in less than five years, the nature of the debate has changed. The Four Rs are now common wisdom, and all serious authorities from virtually all political spectrums agree that land disposal is the worst way to manage our wastes. This change arose largely because of the emergence of the new grassroots movement against toxics, a movement that has brought a new tone and toughness to dealing with environmental threats.

The grassroots movement rejects compromise, taking what Barry Commoner called the "hard path."

> The front line of the battle against chemical pollution is not in Washington but in their own communities. For them, the issues are clearcut.
> . . . In these battles, there is little room for compromise; the corporations are on one side and the people of the community are on the other, challenging the corporation's exclusive power to make decisions that threaten the community's health.

One of the biggest threats to this new consensus on the environment is the sharp difference in approach between the new grassroots environmental activists and the established environmental groups who preceded them. Commoner criticizes the environmental establishment for having

> taken the soft political road of negotiation, compromising with the corporations on the amount of pollution that is acceptable. . . . The national organizations deal with the environmental disease by negotiating about the kind of Band-Aid to apply to it; the community groups deal with the disease by trying to prevent it.

Ken Geiser of Tufts University, in a March 1987 paper, took a look at the *results* that come from the "hard" versus "soft path" approaches:

> In many traditional areas of environmental concern, namely acid rain, pesticides, and wilderness protection, there were frustrating failures in 1986. In contrast, the programs that did win significant victories were all focused on the control and cleanup of toxic chemicals and hazardous waste. Rather than a year of environmental victories, 1986 is better seen as a year in which environmental campaigns around toxic chemical control have succeeded in the face of the continuation of the disappointing environmental lobbying struggles of the 1980s. . . . On close inspection, these toxic chemical campaigns demonstrate a significant difference from the traditional, mainstream environmental movement in terms of constituents and tactics.

Environmentalism's history shows that we succeed when we consciously and systematically focus on building a political base of support within the American public. This holds true for *all* environmental issues, whether they be national parks, endangered species, toxic waste, or garbage. Efforts to preserve and improve the environment are sure to be set back, if not fail outright, when advocates for the environment forget or ignore the fact that environmental causes are just as political as any other public policy issue.

It is not only practical to work deliberately toward building a public base of support for environmental issues, it is also the just and proper thing to do. Most ways of looking at environmental ethics are based on the belief that we all live in an environmental system all of whose components are linked in important and subtle ways. As human beings, we are responsible for exercising proper stewardship of the environment, which means that we must care about human beings and natural resources as well as other forms of life. To ignore any part of the environment is to take an incomplete and fundamentally flawed approach. Efforts for the environment fail when they take people "out of the loop." We believe public support for the environment faded in the 1970s because our colleagues and predecessors either ignored people or saw them as the problem rather than a key part of the solution.

In 1988, Americans will elect a new president. This is good news and bad news. The good news is that all of the likely winners in November have better attitudes toward the environment than Ronald Reagan. The bad news is that believing things will get better under the next admin-istration can be a trap. The last two times we had a positive change in administrations (from Nixon to Ford and from Ford to Carter), the en-vironmental movement backed off, assuming the new administration would do a better job automatically. Both times we were disappointed. And both times the consensus supporting environmental protection was allowed to erode. The real key to meaningful change in national policy on the environment is to continue to build a grassroots consensus for positive environmental policy.

As we move into the next decade, the greatest threat we face is forgetting the lessons of the past. We have learned the hard way already that we must have faith and trust in people's common sense and willingness to act when they understand the issues in terms of right and wrong. We have learned the hard way that law *does not* equal justice. Environmental litigation is useful simply as a way of securing gains won through grassroots political action, which is how good laws get passed. We have learned the hard way that you do not get those good laws without a grassroots political base. Slick lobbying techniques and clever research do not make public policy.

All sorts of approaches at many different levels are useful, but we have learned the hard way that there are only two sources of real power: people and money. Since we will never be able to muster the kind of money-based power the polluters have, we should focus on the most effective ways to build people power.

Love Canal touched something in American society that had not been seen in decades. Hundreds of people contacted Love Canal leaders for advice on how to apply Love Canal tactics to their own local toxics problem. By the end of Love Canal (marked in October 1980 by President Jimmy Carter's buy-out order), there were hundreds of active grassroots toxics groups all across the United States and Canada. By the end of 1987, CCHW counted more than 2,500 groups in every state, Puerto Rico, Canada, and around the world. Even Poland's Solidarity Union adopted toxics as a powerful issue to mobilize the community. For us to go from a virtual standing start to a movement of this size and scope says something about the yearning that exists in American society both for change and for a new way of bringing about that change.

Even though there is always the chance that we will make the same mistakes as those that have gone before, we face the future with optimism. The new grassroots movement against toxics is both broad and deep. Enough attention has been given to building the kind of leadership, as well as public support, that will not let anyone—polluters, government officials, establishment environmentalists, or even CCHW—dictate policy. This time, the new environmental movement is made up of the most basic building blocks, namely, local grassroots groups. Other elements of the environmental movement would do well to reflect on their own specific histories and draw on the lessons those histories teach in the same way we continually examine the lessons of the toxics movement.

In summary, I believe the most important issues we face are waste disposal issues. Why? Because they are concrete and specific "backyard" issues. Since Love Canal, waste disposal issues have made environmentalists out of millions of people who never would have dreamed of going by that title. This has had a wonderful ripple effect with all sorts of other environmental issues.

Some environmentalists have criticized the grassroots toxics movement for being crude and selfish. We are sometimes called NIMBYs (Not In My Back Yard), as though it is a moral defect to fight for home, family, and children. But now, many of the *solutions* the grassroots toxics movement has promoted—waste reduction, recycling, reclamation, and reuse—have become the common wisdom.

I also believe the best way by far to address these issues is to deal with them at the grassroots level. I believe it is morally right that people directly affected by the problems decide how they will be fought. I also believe that the way toxics issues have been fought since Love Canal has been of practical benefit to environmentalism as a whole. As both Barry Commoner and Ken Geiser point out, grassroots environmental activism wins, thereby—as I see it—strengthening the entire environmental cause. That is why I feel my children and their children will have a chance for a better environment in the twenty-first century.

19

Heal the Earth
Heal the Soul

Michael Frome

ON THE DAY IN 1968 that Martin Luther King, Jr. was shot and killed in Memphis, I was at Yale University to speak on conservation policy, certainly including the preservation of wilderness. Once alone following the program, I felt deeply disturbed, trying to equate my actions and personal goals with the tragedy and meaning of Dr. King's life. I asked myself then (and many times since) whether environmentalism and wilderness can be valid in the face of poverty, inequality, and other critical social issues.

The ghetto is a symbol of modern environmental disaster. On one hand, the affluent escape crowds, concrete, and crime by moving to the suburbs. They breathe cleaner air in a cleaner environment. On the other hand, the poor, especially the nonwhite, cannot make it. They are disenfranchised from the bounties of our time. The lower the income, the lower the quality of life, but the higher the air pollution and the diseases from it.

I learned an important lesson in Memphis. I had been there before Dr. King's death and had written about the conservation efforts of a hardy group called the Citizens to Preserve Overton Park. On the face of it, the Citizens had nothing in common with the humble black garbage workers whose cause Dr. King had come to defend. Or perhaps they did, considering they were fighting exactly the same economic and political forces.

Overton at that time had already been a park for almost seventy years. Though less than half the size of Central Park in New York, the woodlands of Overton Park, with seventy-five varieties of trees, are probably richer. It is, in fact, one of the few urban forests left in the world. However, when downtown merchants and developers decided that a freeway through the park would jingle coins in their pockets, the distinctive urban forest became expendable. The two Memphis daily newspapers led the battle for the freeway, belittling any politician who dared stand up in behalf of the park. A former mayor of the city, Watkins Overton, great-grandson of the man for whom the park was named, courageously spoke of the park as hallowed ground—a priceless possession of the people beyond commercial value. Nevertheless, he and the upper side of Memphis learned painfully, along with the garbage workers, that democracy can be "a government of bullies." As Overton said, "Entrenched bureaucracy disdains the voice of the people but eventually the people will be heard."

That idea is paramount in my mind. Entrenched bureaucracy of all kinds disdains the voice of the people. It is the weakness of institutions, whether private or public, profit-making or eleemosynary, academic or professional. Institutions, by their nature, tend to breed conformity and compliance; the older and larger it becomes, the less vision the institution expresses or tolerates. But eventually the people will be heard, as evidenced in the ultimately successful efforts of both the garbage workers and the defenders of Overton Park.

The pioneer ecologist Paul Sears said, "Conservation is a point of view involved with the concept of freedom, human dignity, and the American spirit." Gifford Pinchot expressed the same idea. "The rightful use and purpose of our natural resources," he said, is "to make all the people strong and well, able and wise, well-taught, well-fed . . . full of knowledge and initiative, with equal opportunity for all and special privilege for none." He conceived forestry as the vanguard of a public crusade against control of government by big business. Under his leadership the Forest Service achieved an early reputation for fearlessness in a system then, as now, constipated with bureaucracy, bungling, and timidity. How times have changed! Pinchot stressed the cause of forestry education to train professionals in a social movement; but foresters today are technical people, focused mostly on wood production, trained to see trees as board feet of timber, which is how the Forest Service conducts its business in the public forests.

The National Park Service is not much different. Its personnel may voice concern for ecology as a principle, but scarcely as something practical in critical need of defense. The best defense, at least in my view, is an alert and alarmed public. But national parks personnel are generally inward-oriented and poor communicators. They know the public as visitor numbers, but not as decision makers. Woe unto the parks person who goes to the public with faith or trust in his or her heart. The parks person is a "professional," which is how he or she learned to appreciate the values of ecology in theory, but conformity and compromise in practice.

Students in most academic programs are bred to be partners of the system, not to challenge it. It is part of the nature of institutions in our time. Whether the issue be social justice, peace, public health, poverty, or the environment, all make candidates for study, research, statistics, coursework, documentation, literature, and professional careers, while the poor remain impoverished, environmental quality worsens, and our last remaining shreds of wild, original America are placed in increasing peril.

Martin Luther King, Jr. saw three major evils—racism, poverty, and militarism—and found them integrally linked, one with the other. I see the degraded environment as a fourth major evil, also joined with the others. Environmentalists speak of concern with forests, water, air, soil, fish and wildlife, land use, and use of resources, but these are only symptoms of a sick society that needs to deal more fundamentally with itself.

Presidents and Congresses, one after another, Democratic as well as Republican, have opposed anything but the most niggardly expenditures to educate and house the poor, provide for the aging, rehabilitate the imprisoned and the mentally ill; in the very same fashion they cannot find funds to protect the soil, safeguard the wilderness, or enhance wildlife.

The United States has spent vast sums for so-called security from other nations, while for a fraction of that amount it could have extended humanitarian aid and eliminated the threat of war.

These official actions reflect a system that places a low priority on human values and natural values, a system that needs to reorder priorities while there is still the chance. The compartmentalized approach to life marks the late twentieth century, but the truth is we are all connected. Living that truth begins by recognizing that every human being is rightfully entitled to housing, work, health care, proper nutrition, an adequate income in a habitable environment, and a world at peace.

I find myself turning increasingly to the state of things beyond the wilderness. The nature reserve cannot be decoupled from the society around it. Now I must consider that in the past ten years the population of our prisons has doubled, that we put more people in our prisons than any other "advanced" country, except South Africa and the Soviet Union, and that we have the highest crime rate. Prisons are overcrowded and notoriously inhumane. Most of those found guilty of crimes against society are themselves the victims of society. By this I mean that prison inmates early in their lives suffered child abuse, incest, brutality, and poverty. The poor and uneducated, society's disenfranchised, feel the fury of the justice system, assigned to the worst conditions and the longest terms, while the insiders, like Ivan Boesky and Lynn Nofziger, guilty of connivance, corruption, and theft, get off lightly, serving short sentences in country-club prisons.

In the last eight years, the proportion of nonmilitary spending has been reduced by eight percent, while military spending has more than doubled. With the United States in the lead, the world spends $1.7 million per minute on military forces and equipment—$800 billion per year—or $30,000 per soldier, as compared with $455 per child for education. The United States ranks thirteenth among nations in infant mortality, ninth in literacy, and first in weapons production.

Something is wrong, critically wrong. A society that produces beggars needs restructuring. Martin Luther King, Jr. spoke of the need for compassion. True compassion, he said, involves more than flinging a coin to a beggar. It is not haphazard and superficial. A true revolution of values soon will look uneasily on the glaring contrast of poverty and wealth.

Natural resource professionals ought to be in the lead of the revolution of values. So should the environmental organizations and the people working for them. The problem is that compassion must be at the root of the revolution of values, while compassion, and emotion, are repressed in the training of natural resource professionals and obscured in the management of organizations. Earlier this year I spoke at an environmental conference in Alaska, after which I received a letter from one of the participants. She wrote as follows: "Not once in the ten years I spent studying forestry and land management while getting the Ph.D. did anyone ever speak about ethics."

I am not sure they speak much about ethics in the training of environmental professionals either. Thirty years ago saving the earth was a mission rather than a career. The spirit of the earlier, pre-World War II leaders pervaded the environmental movement, among them Aldo Leopold and Jay "Ding" Darling in the wildlife field; Bob Marshall and Robert Sterling Yard in wilderness and national parks; and Will Dilg, who sparked the organization of the Izaak Walton League. Thirty years ago they were personally remembered and spoken of by those who knew them; Olaus Murie, Benton Mackaye, and Arthur Carhart were still alive. Organizations were led by self-sacrificing missionaries and zealots like Howard Zahniser of the Wilderness Society; Joe Penfold of the Izaak Walton League; Fred Packard and Devereux Butcher of the National Parks Association; and the "archdruid," David Brower of the Sierra Club. Some of their directors, like Sigurd Olson and Harvey Broome, were cut from the same cloth. Broome was a particular friend of mine, a successful lawyer in Knoxville, Tennessee, who gave up his practice to become a law clerk to a judge, with the understanding that he could take time off as needed to pursue his primary interest in wilderness preservation, as a member of the governing council and president of the Wilderness Society.

Now, people go to some very respectable colleges and universities to train for professional careers in the environment. My own daughter attended Williams College and the Kennedy School of Government at Harvard, which I suppose qualifies her for the position she now holds with the Environmental Law Institute. But I am not sure that she experienced the ethical cause of the poor and disenfranchised relative to a healthy environment.

During the sixties and seventies, I observed an abundance of environmental reforms, but reforms are all they proved to be, rather like masking the corpse. In the sixties the Interior Department was busy promoting the establishment of new national parks, which environmentalists cheered, while at the same time underwriting massive power plants in the Southwest, degrading the air quality of the Grand Canyon and other national parks of the region. Interior Secretary Stewart Udall spoke of the good that would come to Hopi and Navajo Indians from exploitation of "their underutilized coal resources," without reference to the inevitable polluted air to which the Indians would be subjected and, even worse, their displacement to make way for strip mining of those "underutilized" resources.

Strip mining was a major issue in the sixties and seventies. Activists in the coal country fought to stop it altogether as heedless destruction of natural and human resources. So did Ken Hechler, a West Virginia scholar in Congress, who pursued principles above political expediency. He insisted that if environmentalists held firm and lobbied hard, they would win. National leaders, however, felt the odds were too long; the best that could be attained was legislation to *control* the practice of strip mining. Thus, I recall being at the White House when President Jimmy Carter ceremoniously signed the new strip mining law. Jubilation prevailed over the prospect of a bright beginning in the coal fields. Essentially, however, that new law legitimized strip mining, establishing standards and a new bureaucracy to enforce them, as best it could, or would, leaving the local activists bitter and disappointed.

The same scenario, with variation, was written in dealing with clear cutting in the national forests. A court decision in the case of the Monongahela National Forest (in West Virginia), based on forest legislation of 1897, was widely interpreted as halting all logging on all national forests, though *selective* logging was still wholly within bounds. In the lobbying confrontations of the 1970s, national environmental leaders, fearing the political power of the timber forestry coalition, opted to accept new systems of planning (via the Forest and Range Renewable Resources Planning Act of 1974 and the National Forest Management Act of 1976), both of which established standards and processes, without seriously curbing clear cutting or reducing the volume of timber removed from the national forests.

Paper victories are tough enough to come by, but they create illusions rather than effective environmental progress. One added illustration involves the national parks. A particular superintendent was assigned by his superiors to prepare a wilderness proposal for his park. The regional director admonished him to leave out two politically controversial portions. "Then," said the director, "we should have no difficulty in getting it through Congress." But the superintendent refused, recalling later, "I was more concerned with saving the wilderness than in getting a law passed saying that I had done so." Or to quote S. Herbert Evison, who served for many years as an official of the National Park Service: "We who have tried to improve conditions have been too ready to compromise. Experience shows it is better to make demands than deals."

Gifford Pinchot may have said it best. In 1910, when forestry was a vital, progressive force in the forefront of the conservation crusade, he wrote:

> We have allowed the great corporations to occupy with their own men the strategic points in business, in social and political life. It is our fault more than theirs. We have allowed it when we could have stopped it. Too often we have seemed to forget that a man in public life can no more serve both the special interests and the people than he can serve God and Mammon. There is no reason why the American people should not take into their hands again the full political power which is theirs by right, and which they exercised before the special interests began to nullify the will of the majority.

The desire for a more environmentally-based society has become deeply rooted since the sixties. The public has shown a high level of support for environmental protection. Despite energy shortages in the seventies, recessions, the cost of environmental laws, and Reagan landslides, a variety of surveys shows little evidence that the public wants to reduce environmental protection programs by less regulation or less spending. Given the choice, a majority favors *less* economic growth. Surveys consistently indicate that people feel protecting the environment is more important than keeping prices down on products they buy.

However, something new must be done by a new group of people. The priority item on the agenda, as I see it, is for those who hope to heal the earth to join with those who hope to heal the souls of our fellows to bring something new to bear. We must face the twenty-first century with new emphasis on human care and concern. A fair profit may be defensible, but profiteering has skyrocketed at the expense of social and environmental responsibility. The proposal to open the Arctic National Wildlife Refuge to oil exploration and drilling offers a classic case in point: It does not have a damn thing to do with meeting human needs. Profiteering should never be glorified, nor confused with social services.

We cannot set aside a little bit of wilderness and say, "That much will take care of the soul side of America." We must rescue everything that still remains wild and recapture a lot more than has been lost, looking to its future rather than its past. In the battle for wilderness there are no enemies. The children of the poor will become rich for what is saved; the children of the rich will be impoverished for what is not saved. It takes considerable courage to stand up against money and the power of politics and institutions. It takes wisdom, or at least knowledge and courage, to work through the system. When the Pope visited the United States he said, "We need more than social reformers; we need saints." I would say, "We need more than social reformers; we need revolutionaries—not to commit violent acts but to press society to reorder its priorities."

"New opinions are always suspected, and usually opposed," wrote John Locke more than three centuries ago, "without any other reason but because they are not already common." Such is the way of institutions, but not of individuals. Only the individualist can succeed, even in our age of stereotypes, for true success comes only from within. When we look at the revolutionary task of reordering priorities, and the sheer power of entrenched, interlocked institutions, the challenge may seem utterly impossible. Yet, individuals working together, or even alone, at the grassroots of America, *have* worked miracles. The odds in Selma and Montgomery, Alabama, also looked impossible, in the long fight for the Wilderness Act, and for Overton Park and many other places like it. "A nation that continues year after year to spend more money on military defense than on programs of social uplift is approaching spiritual death," wrote Martin Luther King, Jr. who embodied in his own self the challenge to spiritual life. Each individual must realize the power of his and her own life and never sell it short. In setting the agenda for tomorrow, miracles large and small are within reach.

20

The National Forests and the Environmental Movement

Randal O'Toole

ENVIRONMENTALISTS HAVE ACHIEVED much success over the past twenty years, yet have fallen far short of their goals. The causes of these shortfalls are usually attributed to shortsighted decision makers and the political power of environmental opponents. But often, it is the environmentalists who are to blame for their incorrect diagnoses of the problems. A prime example of this is the debate over national forest management by the USDA Forest Service.

The 191 million acres of national forests are both the glory and the bane of the environmental movement. Most of the forests were created in the 1890s and 1900s and are today considered one of the greatest legacies of the progressive conservation era. But increasing timber cutting that began in the 1950s caused major public controversies and contributed to the creation of the modern environmental movement.

Gifford Pinchot, the dynamic leader of the Progressive movement, founded the Forest Service with the idea that scientifically trained foresters would manage the national forests in the public interest. One of the most outstanding such foresters was Aldo Leopold, who—together with other Forest Service professionals—literally invented two important concepts: wilderness and wildlife management.

As forest supervisor of New Mexico's Gila National Forest, Leopold approved the designation of the nation's first wilderness area in 1924. With the support of top Forest Service officials, the next twenty years saw the dedication of millions of acres of land to the wilderness concept. Leopold is also credited with being the father of wildlife management. He was among the first to recognize the importance of habitat to wildlife and later founded the first school of wildlife management at the University of Wisconsin.

The philosopher-king idea of forest management became less enchanting during the 1950s. Depletion of private timber supplies combined with postwar housing demands led to an increasing emphasis on timber management in the national forests. Clearcutting, herbicide spraying, and the declassification of wildlife areas to access their timber aroused public opposition in every region of the country. Debates over these issues have continued to this day.

Believing that the Forest Service is naturally biased toward timber, most environmentalists conclude that prescriptive legislation in some form or another is necessary to resolve these issues. Led by the Sierra Club, opposition to Forest Service policies focused on several demands: congressional designation of wilderness areas to protect them from the Forest Service, a ban on or great reduction of clearcutting in the national forests, protection for old-growth and/or the management of timber on very long rotations, protection of important watersheds, and maintenance of wildlife habitat.

With respect to the national forests, the environmental movement has scored two great victories: the Wilderness Act and the Endangered Species Act. Since passage of the Wilderness Act in 1964, millions of acres of national forest lands have received congressional protection. The Endangered Species Act of 1972 requires the Forest Service and other federal agencies to protect numerous species of birds, mammals, and other wildlife by modifying management practices.

There is no question that the Wilderness Act has succeeded in protecting large amounts of land from timber cutting and other development. Yet, millions of acres of lands that deserve wilderness status are likely to be developed by the Forest Service for timber sales—mostly at an economic loss. For example, the Forest Service plans to construct roads and log 1.2 million acres of roadless lands in Montana national forests. Such development serves no purpose other than timber harvesting, yet nearly all of the timber sales will lose money. Wilderness bills now being considered by Congress make only a small dent in these proposals.

The Endangered Species Act is not as clearly successful as the Wilderness Act. Whereas some projects have been halted or modified to account for rare species, many believe that the Fish and Wildlife Service, which administers the act, places undue emphasis on individual species rather than habitat protection. Captive breeding programs for the California condor, black-footed ferret, and other species are an expensive and inadequate substitute for sound habitat.

The Yellowstone grizzly bear is a particular example of how the Endangered Species Act has failed to influence national forest management. The bear unquestionably needs solitude more than anything else that people can provide. Yet, plans for forests surrounding Yellowstone National Park call for road construction and timber harvest in hundreds of thousands of acres of prime grizzly habitat. The fact that such timber sales are almost all below cost underscores the misplaced priorities of the Forest Service.

Other than these two laws, the environmental movement has failed to convince Congress of the need for prescriptive legislation regulating national forest management. Congress specifically rejected prescriptive proposals regarding such issues as clearcutting and the age of timber harvest when it passed the National Forest Management Act (NFMA) in 1976. The few prescriptions included in NFMA were already standard Forest Service policy and were included with the approval of the agency.

In any case, the historic record does not indicate that legislation prescribing forest practices will be successful. A federal court ruled in 1974 that the Forest Service was violating an 1897 law forbidding clearcutting. Such violations continued until the law was repealed in 1976. A 1976 court ruling found that a 1904 law prohibiting public entry into the municipal watershed for the City of Portland was violated by large-scale Forest Service timber cutting starting in 1958. That law was repealed in 1977.

A 1972 law limiting timber harvesting in the Hells Canyon National Recreation Area to "selective cutting" (uneven-aged management) was overturned administratively by orders from the Reagan administration to use shelterwood cutting, which, like clearcutting, is even-aged management. Lengthy and expensive court battles can be fought against such actions, but, as the first two experiences suggest, success in the courtroom can be followed by repeal in Congress.

As a substitute for prescriptive legislation, Congress decided that a comprehensive land and resource management planning process should be incorporated into the National Forest Management Act. Accompanied by extensive public involvement, this process was supposed to resolve disputes over clearcutting, wilderness, and other issues. If anything, forest planning has increased polarization and conflicts over national forest management. Most plans have ignored the major issues addressed by NFMA.

For example, although clearcutting was the problem that led to NFMA, few of the 118 forest plans published to date have considered seriously alternatives to even-aged management. Concerns over below-cost timber sales, roadless areas, herbicides, overgrazing, and other issues have been ignored in most if not all plans. In most instances, the plans are little more than justification statements affirming the status quo.

One unintended benefit of the planning process has been the environmentalists' growing mastery of technical and especially economic matters. In 1975, debates over national forest management were characterized by shrill emotionalism on the environmental side in contrast to apparently reasoned scientific analysis on the industry side. Today, the reverse is true, with industry appealing solely to the jobs issue while environmentalists have the support of biological, physical, and economic sciences.

This technical sophistication is changing the standard environmental paradigm. Traditionally, environmentalists have believed that national forest problems are due to the timber bias of Forest Service officials. Recent economic research and analyses suggest that these assumptions are false and that alternative approaches to reforming the Forest Service may be more successful than the traditional environmental goal of prescriptive legislation.

For example, reviews of forest plans have shown that the vast majority of national forests lose money on timber management. When returns to the Treasury are compared to Treasury costs, even forests with valuable old-growth timber often lose money. Yet, a close analysis of below-cost sales shows that they are due not to a Forest Service bias toward timber but to the agency's tendency to maximize its budget.

One old-growth forest that loses money is the Tahoe National Forest in northern California. This forest collected more than $17 million in timber receipts in 1987. This appears impressive because the Forest Service spent only about $4.3 million in tax dollars for timber sales, timber management, and timber-related road construction. Of the receipts, however, $2 million were "paid" by the timber purchaser in the form of road construction. Purchasers are allowed to credit the cost of road construction against the price they bid for timber. An additional $9 million were retained by the Forest Service for postharvest management activities, such as fuel treatment, reforestation, and herbicide spraying. Counties are paid about 25 percent of timber receipts in lieu of taxes, so they received about $4 million. Therefore, the return to the Treasury was only about $2 million. When this is compared with the $4.3 million cost, the Treasury lost more than $2 million.

The Forest Service often claims that such losses are justified by the benefits timber sales provide to other resources, such as roads for recreation and openings for wildlife, and by the need to increase productivity by harvesting slow-growing trees and reforesting with young trees. Forest planning documents reveal, however, that such claims are specious.

Almost all national forest plans indicate that existing forest roads provide more than enough roaded recreation to meet current and future recreation demands. New roads may be used by recreationists, but this use would merely be transferred from another part of the forest; no new value would be created. In most national forest, roadless recreation already is or will be in short supply long before roaded recreation opportunities are used to capacity.

Some wildlife species such as deer and elk benefit from openings that provide a source of food. But most national forests already have sufficient openings, and cover for protection from predators and temperature extremes is more important. Researchers have found that elk avoid timber-related roads in the Rocky Mountains and that deer require old-growth forest in Alaska for food and protection from heavy snows.

Investments that the Forest Service plans to make in growing new trees merely will produce a new crop of below-cost timber. Economic analyses indicate that reforestation on most national forests will return as little as 10¢ to 50¢ for each dollar invested. Below-cost timber sales certainly are not justified by such returns.

Close scrutiny of Forest Service operations shows that below-cost sales are the predictable result of a number of well-intended but poorly designed laws that give managers an incentive to lose money on timber. The most important of these laws is the Knutson-Vandenberg Act of 1930, which allows the Forest Service to retain funds out of timber sale receipts for reforestation. In 1976, Congress expanded the use of K-V funds to include precommercial thinning, trail construction, and other so-called forest improvements, giving managers a strong incentive to sell timber. Because the cost of arranging timber sales and building timber-related roads is borne by taxpayers, the Forest Service makes no effort to ensure that timber receipts cover these costs. It does require, however, that a minimum bid prices include "necessary" K-V funds. In many timber sales, almost all receipts are placed in the K-V and similar funds, leaving little or none for the Treasury. Taxpayers lost money on the Tahoe National Forest in 1987 because more than half of the forest's timber receipts went into the K-V and related funds.

In 1987, the Forest Service retained more than $300 million out of timber receipts for reforestation and other activities. This is more than the total that Congress appropriated for recreation, watershed, fish and wildlife, and range management combined. Managers see an immediate reward for timber sales because about 75 percent of the timber receipts retained by the Forest Service is spent on the district that earned the receipts. All levels of the bureaucracy encourage timber sales because the remaining 25 percent is spent on "overhead" by the Washington, regional, and forest supervisors' offices. Forest Service specialists in watershed, fish and wildlife, and recreation often support timber sales because K-V funds can be spent on these activities.

The suggestion that Forest Service officials use K-V funds to increase their budgets does not imply that officials are unscrupulous or unethical.

Most national forest officials are truly interested in doing what they consider to be best for the land, but through a process of natural selection those managers who believe in activities that increase the agency's budget are the ones who are promoted. After several decades, the Forest Service has become dominated by people who truly, but mistakenly, believe that timber cutting is good for recreation, wildlife, watershed, and other forest resources.

The idea that budgets are a primary influence on Forest Service managers casts national forest problems in a new light. Gifford Pinchot's hope that scientifically trained foresters would manage the national forests in the public interest with no regard for their own self-interest has no foundation. This is not, as many environmentalists have thought, because Forest Service officials are by training or nature biased toward timber. Instead, agency employees are just like anyone else: interested in their paychecks, prestige, authority, and things they can accomplish. All of these things are increased with larger budgets. As even the most altruistic managers need a budget, factors influencing budgets are important incentives for agencies, including the Forest Service.

A review of national forest controversies, including wilderness, herbicides, and below-cost sales and grazing, indicates that most are rooted in misincentives created by the Knutson-Vandenberg Act and similar laws. Forest managers can increase their budgets by selling timber, not by creating wilderness, so naturally they will oppose wilderness. Herbicides are funded by K-V collections, and herbicide spraying results in overhead paid to higher offices, so officials tend to support herbicides. Just as K-V funds give managers a powerful incentive to sell timber below cost, the Range Betterment Fund, which gives managers half of all grazing receipts, provides an incentive to lease grazing rights below cost.

This analysis is new to environmental thinking. When the blame for overcutting and overgrazing is laid on managers who are somehow biased toward commodities, prescriptive legislation appears to be the best solution. But if these problems are caused by misincentives created by poorly designed laws such as the Knutson-Vandenberg Act, a completely different solution is required.

This solution is to change the incentives that govern forest management. First, the Knutson-Vandenberg Act and related laws that give managers an incentive to harvest timber and graze domestic animals below cost must be repealed. Instead, timber and grazing should be funded out of a fixed share—perhaps two thirds—of net receipts. Then managers will have an incentive to sell timber only where timber sales make money, because below-cost sales will reduce net receipts, thereby reducing the Forest Service's budget.

Because most forests lose money on both timber and grazing, funding activities out of net receipts will reduce environmental conflicts greatly. But some forests, particularly those in the Pacific Northwest and the deep South, make money on timber management, and many other forests have at least some land that can make money. It is not enough simply to remove the incentive for below-cost activities; managers also must be given an incentive to protect recreation, wildlife, and other amenity resources.

One of the best ways to do this is to allow forest managers to charge fees for recreation and give them the same two-thirds share of net recreation receipts. This would more than double funding for recreation and add millions of dollars to the annual budget of almost all national forests. To protect this income, forest managers would have an incentive to maintain scenic vistas, improve wildlife habitat, protect water quality, and otherwise moderate management of above-cost timber. To enforce collection of fees, recreationists would be required to display on their car or person a visible permit, much like a parking permit or ski-lift ticket, that indicates they have paid an annual or short-term fee. Experiences with other forms of recreation indicate that such a system would have a low cost and high degree of compliance.

Not all problems can be corrected by changing national forest incentives. Some rare or endangered species such as the grizzly bear will benefit from an end to below-cost sales, but others such as the spotted owl still will require the protection of the Endangered Species Act. Recreation fees provide some incentive to protect water quality, but much national forest water is consumed off the forests, so water quality will continue to require the protection of the Clear Water Act.

Yet, environmentalists' increasing awareness of incentives is likely to change the character of national forest debates over the next few years. Traditional opponents of prescriptive environmental legislation will tend to support changes in national forest incentives. Fiscal conservatives, for example, will support an end to below-cost sales. Counties, which receive 25 percent of national forest receipts and tend to support industry demands for more timber sales, will support instead both recreation fees and an end to economically marginal sales, as these actions will increase county receipts.

The carrot works better than the stick. New incentives for national forest managers will protect many resources better than prescriptive legislation. With support from counties, fiscal conservatives, and other traditional opponents of prescriptive legislation, changing national forest incentives is also more politically feasible than is prescriptive legislation. Improved incentives will resolve numerous environmental conflicts so environmentalists can concentrate their energies on those that remain, such as Pacific Northwest old growth. New incentives could also be applied to the Bureau of Land Management, Corps of Engineers, and other agencies that spend tax dollars on environmentally destructive projects.

Conservationists can take credit for the creation of the national forests and environmentalists can credit themselves for legislative protection of millions of acres of wilderness. But further progress toward protection of public resources from government-subsidized commodity exploitation will require a new direction from environmental leaders. That direction can be found in the area of incentives.

21

In a Landscape of Hope

Charles E. Little

IN 1933, A FORESTRY PROFESSOR named Aldo Leopold published an article in the *Journal of Forestry* setting forth a proposition that today has become something of a growth industry amongst the intellectuals of the out-of-doors: the land ethic. "That land is a community," said Leopold in a later description of the idea, "is the basic concept of ecology, but that land is to be loved and respected is an extension of ethics." Professor Leopold's idea was published again as the last chapter of *A Sand County Almanac* in 1949, the year after he died fighting a brush fire near his summer home in Wisconsin. The book lives on and through many editions has sold in the millions, especially to the "greening of America" youth who are now professors themselves.

Leopold suggested caution and deferred rewards in our use of land resources as an ethical proposition, rather than an economic one. Land is a community, he said, not a commodity. We would do well to use it with love and respect. In time Aldo Leopold became the most quoted author in conservation circles with the possible exception of Thoreau.

The idea of ethical evolution, as Leopold expressed it, is simple enough. The article opens with the tale of Odysseus returning from the Trojan wars and hanging "all on one rope a dozen slave-girls of his household whom he suspected of misbehavior." Leopold states that for Odysseus this was simply a matter of the disposal of property, not of right and wrong. The ethic—which is a *social* concept—of not summarily executing the girls below stairs had not yet caught up to history. After a while, perhaps as the Christian concept of equal souls before God emerged, it did become antisocial to kill one's servants in such a manner. And in further developments, society extended ethical behavior to cover quite complex human relationships—between and among families, tribes, communities, even nation-states. But what about the land?

Our currently sophisticated extension of social ethics reaches a variety of institutional relationships, but not, finally, the "land-relationship" as Leopold called it in 1933. "There is as yet," he wrote, "no ethic dealing with man's relation to land. . . . Land, like Odysseus' slave-girls, is still property. The land-relation is still strictly economic, entailing privileges but not obligations."

Clearly, the land ethic, if it is ever to be truly expressed, will have to be an ethic not to serve the political economy but to serve the land whether it is economic or not: an ethic that advances the *land's* reasons for being. This is what Aldo Leopold would teach us, and I believe it. But it's hardly the case at present any more than it was in 1933. In fact, as far as land is concerned, the present is a flop. If you do not believe this, go home again.

I did.

At the foot of the San Gabriel's talus slope, where the vineyards (and orange groves and truck farms) fructified in the 1930s, there had been a narrow, two-lane concrete highway. A canopy of fragrant eucalyptus trees had shaded the road as it meandered through the valley, and shaded, too, the long strung-out column of tramps looking to do chores for food in those days when there was no money. They came to our settlement along California State Route 118, a wide place where there was a grocery, and a post office, and Jack's gas station, and even a tiny library. In springtime the tramps would come through the gate of our bright-green picket fence to sit on the back steps of our old house, a block off the highway, and eat sandwiches my mother had prepared for them. I would watch them from my perch on a low limb of the great pepper tree that shaded the yard, and they would say, "Howdy, boy," to me, like Woody Guthrie. And, "You some tow-head, you are," in a thick drawl from Oklahoma or Arkansas or the Texas Panhandle, and then they'd tip their hats and say, "Thank you Ma'am," to my mother and wink at me and be on their way to the next little town along the highway where produce trucks and flivvers stuffed with furniture and children and hope chugged along in some dance choreographed by the economics of the land in those years.

The hope was the amazing thing. And the faith: faith that new land could be found that would not wear out and turn to dust.

Eventually, I found our old house. For I wanted hope too. But the bungalow's stucco was now mottled and flaky, the pepper tree gone, the picket fence with it. The house itself was standing precariously on the edge of a cliff—a cliff of concrete surmounting the freeway, the new six-laner which now cut through my valley.

I looked down on the three eastbound lanes where, I think, Mr. Lee's place was, a small chicken ranch now hovering in memory about twenty feet above the streaming traffic. I wondered where all this machinery was going, and had a vision of the great river of cars disappearing at the edge of the earth after a million miles of shopping centers, eroded fields, pastures grown to brush, suburban-kitsch office buildings, clear-felled forests, drive-in banks, dammed rivers, muddy lakes, festoons of high-tension wires, and Wendy's and Hardee's and Arby's disappearing into a taupe-colored distance.

Mr. Lee had always told us not to put our finger through the chicken wire lest the Leghorn rooster, who was cranky, come peck at it. And so we would put our finger through the chicken wire, and rooster would come and peck it, and we would yowl, and Mrs. Lee would give us a cookie.

Aldo Leopold wrote when I was two: "To build a better motor we tap the uppermost powers of the human brain; to build a better countryside we throw dice."

Zoom zoom zoom go the cars along the freeway. We put our fingers through the wire. We roll the dice. The land disappears. Mrs. Lee is no longer there to give us a cookie. And Mr. Leopold is dead.

Is there a "land ethic" in heaven? I hope so for Aldo Leopold's sake, because he despaired of there being one here on earth. He despaired in the first published version of the idea, and the despair was unallayed by the passage of years. In the final version of *A Sand County Almanac,* written after the war and the atom bomb, Leopold bitterly concluded that "no important change in ethics was ever accomplished without an internal change in our intellectual emphasis, loyalties, affections, and convictions. The proof that conservation has not yet touched these foundations of conduct lies in the fact that philosophy and religion have not yet heard of it."

And that is our text for today: because the fact is that philosophy and religion *did,* eventually, hear of conservation and of the land ethic that is its philosophical cornerstone. These days, the concept is no stranger to either podium or pulpit. And therefore might *we* be justified in predicting a different future for the land than could Professor Leopold? Despite the blasted landscapes of the present, might not ethical considerations finally be set into the grain of our future public and private decisions about land use and conservation? Or shall we keep on tossing the dice, again and again, until the land craps out?

If you are looking for hope, please attend to the words of the U.S. Catholic Bishops in their November 1984 pastoral letter, "Catholic Social Teaching and the U.S. Economy."

> The biblical vision of creation has provided one of the most enduring legacies of church teaching, especially in the patristic period. We find a constant affirmation that the goods of this earth are common property and that men and women are summoned to faithful stewardship rather than to selfish appropriation or exploitation of what was destined for all. Cyprian writes in the middle of the third century that "whatever belongs to God belongs to all," and Ambrose states "God has ordered all things to be produced so that there would be food in common possession of all." Clement of Alexandria grounds the communality of possession not only in creation but in the incarnation since "it is God himself who has brought our race to communion (*koinonia*) by sharing himself, first of all, and by sending his word to all alike and by making all things for all. Therefore everything is in common."

> Recent church teaching, as voiced by John Paul II, while reaffirming
> the right to private property, clearly states that Christian tradition "has
> always understood this right within the broader context of the right
> common to all to use the goods of the whole creation."

Applied to land, this is as clear an ethical pronouncement as one
could wish. To some it is shockingly clear. At a meeting held to discuss
the implications of the Bishops' letter for managing land resources, a
government economist confessed his dismay. "I have a Ph.D. in economics,"
he said. "And in all my studies, I have never seen as radical an economic
document as this."

You see, we commonly take land to be, mainly, an economic "input":
with labor and capital, a "factor of production." The Great Plains are an
input into the agriculture industry. The timbered Northwest is an input
to the forest products industry. The wilderness fastness is an input to the
recreation industry. And my valley was an input to the real estate industry.
Leopold made this curiosity familiar, describing it as a kind of resource
Babbittry. I personally know people who go around muttering "land is a
factor of production" all day long, without even realizing what they are
saying, just as some of us say grace at dinner, "God is great, God is good/
And we thank Him for this food," without wondering who *really* owns
the land that makes the food. And the table. And the china and silver and
tablecloth, too. At this level of inquiry, it's hard to understand land ownership in any but the most transient and inconsequential sense. Ownership:
This is the linchpin in the whole business of land ethics, of course, as the
Bishops so forthrightly assert.

A real philosopher I know (Sara Ebenreck, who has a Ph.D. in ethics
from Fordham) tells of the young chief of the Western Cayuses who in
1855 protested the selling of the tribal lands. "I wonder if the ground has
anything to say," he asked the governor of the Washington Territory. "I
wonder if the ground is listening to what is being said?"

Owning land—in the monopolistic, exploitative sense, not in Jefferson's sense that all should be allowed "a little portion"—has always seemed
a bit like owning the air through which we pass, or the waters that fall
or flow or tidally undulate. Land *moves,* like air and water. And we move
through it in our brief lives. It opens before us and it is well to wonder,
after we have passed: Do other travelers and voyagers find it good? What

does the ground say? I am often astonished when people talk about the need for a land ethic as if it were an argument about table manners. It is not. It is an argument about violence, as Leopold made plain in the very first lines of his essay. The stewardship of land is a form of not raping it.

"How do you feel about not-raping?" asks the fellow next to us at the cocktail party, for he has somehow discovered that we are the holder of strange views.

"Well, I'm all for it," I guess we are supposed to say. "There ought to be a whole lot more not-raping going on. We got to get the word out."

"Still," says the fellow at the cocktail party, fingers glistening with chicken grease from the barbequed wings of a factory-made Leghorn, "you can overdo the idea of not-raping. After all, we have to be practical. This is a free country. A man has his rights. I'm sick and tired of all those do-gooders running around complaining all the time. Let's stand up for America."

But we are called to stand up for the land, too.

Another, earlier, statement by the Catholic Bishops—those whose sees are in the American heartland—asserted that the Bible and the tradition of the church make manifest these ten principles of land stewardship:

1. The land is God's.
2. People are God's stewards on the land.
3. The land's benefits are for everyone.
4. The land should be distributed equally.
5. The land should be conserved and restored.
6. Land use planning must consider social and environmental impacts.
7. Land use should be appropriate to land quality.
8. The land should provide a moderate livelihood.
9. The land's workers should be able to become the land's owners.
10. The land's mineral wealth should be shared.

An ecumenical group of North Carolina churchmen called the Land Stewardship Council writes in its "Ministry Statement" of 1981:

We are all Creatures of God. We and the land are the work of God's creative love. The strong basis for the traditional Jewish-Christian

272

concept of stewardship can be seen in numerous places in the Scrip-
tures. The Bible describes the proper relationship that people should
have with the land and with each other. This is expressed plainly, for
example, in Psalms 24:1—"The earth is the Lord's and all that is in
it, the world and all that dwell therein." In Leviticus 25:23, God says
that no land should be sold in perpetuity "because the land is mine";
to me you are "aliens and settlers."

In Minnesota, an outfit called the Land Stewardship Project has created its own bible. Put together by Joe and Nancy Paddock and Carol Bly,
poets and writers, the book is a compendium of long and short quotations
by other poets and writers interspersed with the editors' own insights. (A
version of the compendium was published in 1986 by Sierra Club Books
under the title *Soil and Survival*.) Black Elk, the Oglala Sioux holy man,
is here. And Isaiah. And E. F. Schumacher. And Walt Whitman. And scores
more you haven't heard of, and don't need to.

Joe Paddock, in his poem, "Black Wind":

This vast
prairie, its hide of sod
stripped back, black
living flesh of earth
exposed.
Our way
has made thieves
of the wind and rain.
Listen,
listen to the wind moan
through the bone-white dead
cottonwood limb: *Half gone!*
Half gone! Half gone! . . .

But half remains, too.

We are, like the Chinese (from Confucius to Mao), a nation in love
with axioms. Our homes and offices are littered with them: from "Be It
Ever So Humble . . ." to "The Buck Stops Here." We wear them on our
T-shirts ("A Woman's Place is in the House . . . and the Senate") and the
bumpers of our automobiles ("Thank You For Not Laughing At My Car").
My own grandfather, a printer, used one on a magazine he published to

promote his business. Under a lithographed team of horses straining at the plow were the words, "Work, Son of Adam, and Forget It."

But axioms are not ethics. Ethics, it seems to me, is work: the work of a society trying to live up to its beliefs. A land ethic proposes restraint in land use, deferred reward from exploitation of the resource base, concern for posterity so that future generations will get as much or more from the land as we. It is a social goal, this land ethic of Aldo Leopold, and it must be expressed in "policy." And not only the abstract, big-P Policy of Principle, but the workaday little-p policy of legislation, of statute, of government regulation and management practice.

While the land ethic has recently laid claim to our consciousness to a degree that might have heartened Professor Leopold, there are other social goals that tend to complicate the effort to create and implement the legislation needed to make it actual. Some of these goals—individual liberty, social justice, scientific progress—are much on the lips of those whose economic ox would be gored by the actual application of a land ethic in policy. They insist that the goals they espouse are in conflict with a land ethic. In fact, so persuasive have the opponents of a land ethic been with this tactic that of all industrial democracies, the United States (which has the most to gain from it) has the least effective legislation to protect its land base.

Whether the arguments are in opposition to wilderness designation, establishing wild and scenic rivers, conserving soil and water resources, planning for urban development, or limiting the conversion of prime farmland, the exploiters of land are adroit at using the rhetoric of the social reformers of yesteryear.

John Locke (1632–1704), who gave us the outline for a liberal constitutional government, provides the most relevant example. Locke proposed the concept of "natural rights"—these being life, liberty, and property.* They are natural because they would inhere to mankind in a "state

---

*In the Declaration of Independence, Jefferson's "inalienable" rights are "life, liberty, and the pursuit of happiness." Jefferson believed that property, including ownership of land, was a right, and he championed it in every respect. As Garry Wills points out, however, the word "property" did not survive in the Declaration because Jefferson, influenced by the Scottish philosopher Francis Hutcheson, felt that property "*follows*" on society rather than precedes it. Thus he places it among the 'adventitious' rights rather than the 'natural'

of nature." He said that government was valid only with the consent of the governed and that it was the "natural right" of men to "dispose of their persons and possessions as they think fit." We now listen to modern day philosophers of the political right asserting their own anarchic version of Lockean liberalism, as in, "It's my land and I can do with it what I want."

Locke, an urbane Londoner, was, in effect, the originator of "individualism" of the kind that is now thought to be a uniquely American characteristic. In its most simplistic form, the American individualist is contemporaneously embodied by the Marlboro Man who rides the plains alone and inhales deeply despite warnings by the surgeon general. Only slightly more subtle are the landowners who believe that moral responsibility stops at property lines.

How does the Lockean individualist view, American-style, comport with the Bishops' communitarian philosophy of land use? The answer is, not very well. And here is the first of several conundrums that arise when we wish to apply the land ethic to policy. It is deepened, at least so it would seem on the surface (and we shall return to this point later), by Thomas Jefferson's small-d democratic insistence on the individual right of land ownership to provide for one's own welfare and subsistence.

Thus are we caught in a trap of our own manufacture. By appealing to authority without sensitivity to the historical setting in which reform-minded concepts to which we mindlessly cling were created, we allow the moral teachings of the past to be perverted by those who would use them cynically. According to Eugene C. Hargrove, a professor of philosophy at the University of New Mexico, a landowner cannot honestly justify his position that he is absolved of social responsibility by asserting a natural right of land ownership to do with his land whatever he might choose. This is a claim, says Hargrove in the Summer 1980 issue of *Environmental Ethics,* that neither Locke nor Jefferson would have been comfortable with given present-day circumstances in which the perverse exploitation of land is exacerbated by limitations on its quantity. Both men thought of the American frontier as virtually endless.

Locke and (less often) Jefferson are not the only authorities patri-

---

ones." See *Inventing America* (New York: Random House, Vintage Books Edition, 1979), pp. 229–239.

otically invoked in defense of unethical land use. Jeremy Bentham (1748–1832), another reformer, proposed that the basis for all legislation was "the greatest happiness for the greatest number." At one time, this radical thought stood in contrast to policies that benefited only the nobility in England. Today, in America, Benthamite utilitarianism is used to justify everything from ski lifts in national forests to governmental sponsorship of the use of poisons in agriculture. Another reformer, Adam Smith (1723–1790), hoped to benefit the masses with his theories expounded in the *Wealth of Nations*. Here it is written that if individuals undertake their "industry" primarily for their own gain, then they will benefit society by "an invisible hand" that frequently produces a better result in serving society's needs than would governmental intervention in order to improve the public welfare. Ever since, industrialists and others have taken the work of the great Scottish economist as license for greed and antisocial behavior in general, and specifically to excuse a failure to think of land resources in terms other than immediate gain from exploitation.

During the 1980s the Free Market has had its best run since Coolidge, "trickle down" no longer a cause for sniggering by limousine liberals. Adam Smith, in an agrarian age, said that government could not do very much to affect the welfare of individuals in society. And if a twentieth-century John Maynard Keynes proved him wrong, and if a fifty-year history of just the opposite created the wealthiest and strongest, as well as the most decent, nation on earth, then so what?

If you want a course in selective Anglo-American intellectual history, all you need to do is attend a hearing—in Congress or at Town Hall—on any legislation or ordinance designed to protect the land resource base, on policies that would give substance to the *idea* of a land ethic, which while much on the lips is scarcely on the books.

In such sacred places as these you will hear Mr. Bentham, who wants to construct a theme park in the last remaining unspoiled marsh in the state's coastal zone, insisting, "Listen, I'm a *people* person." Or Mr. Locke, who wants to build a 3,000-unit townhouse development on some prime farmland, complaining, "I don't need a bunch of conservationists telling me what I can do with my land." Or Mr. Smith, whose nuclear power plant is to be sited atop the local fault line, crying, "Jobs, jobs, jobs!"

They are, each of them, historically correct, having got hold of some solid philosophical precepts. But they are tragically wrong, too. And the dilemma doesn't bode well for the future of the land ethic.

What do we build on then? Most important are the citizen-effort models, of course: the heartening case histories of those who earnestly try to express the land ethic in terms of civic action in small as well as large ways. Perhaps the small ways are better for purposes of inspiration. It is one thing to fight hard for the highly visible conservation goal, such as, say, the protection of the Alaskan wilderness. But it takes nothing away from that achievement to remember that smaller victories (albeit equivalently partial) may be even more expressive of the internal change of intellectual emphasis, loyalties, affections, and convictions that Aldo Leopold said was the sine qua non of a functioning land ethic.

For example, a land ethicist I know, Tom Lamm, who works out of Black Earth, Wisconsin, is helping to organize small farmers to do, finally, what politicos from FDR on down have been afraid to do: make soil erosion against the law. A recent account of the work of the Soil Stewardship Task Force in the *Wisconsin State Journal* quotes the farmers who make up the task force as saying that "Regulations must be set in place to control abusive soil eroders who have not, and will not, respond to technical assistance and financial incentives alone." It is the small farmers, Lamm believes, who have the largest sense of land stewardship and who must therefore take the lead in making laws about the care of the land. Otherwise its future is left to real estate investors and other absentee owners to whom land is mainly surface and not the magical thing that a real dirt farmer knows it to be.

There are a good many people like Tom Lamm who give us hope and inspiration. The fact is, I have a Rolodex full of them. Eddie Albert continues to celebrate "green acres" in soil erosion work. Ned Ames gives away money for a foundation to preserve natural areas. George Anthan writes articles for a major midwestern newspaper on farmland preservation. Malcolm Baldwin, when at the President's Council on Environmental Quality, rescued from certain oblivion the Agricultural Lands Study, which alerted the nation to the loss of farmland. John Banta guards the legal standing of the Adirondack Park Agency, which protects some six million wilderness and semiwilderness acres in New York State. Others in the B section are land ethicists too: Batcher, Beale, Beamish, Beard,

Beaton, Beaty, Becker, Berg, Berger, Bergland, Berrett, Berry, Bodovitz, Boon, Borgers, Borrelli, Boswell-Thomas, Bray, Brinkley, Brooks, Brown, Browne, Burch, and Burr. And it goes on like that through Zinn, Zitzmann, and Zube. Zinn is a geographer who works for the Congressional Research Service and edits a newsletter on coastal zone resource management. Zitzmann is a recently retired land use planner from the Soil Conservation Service. Zube, at the University of Arizona, is a leading figure in landscape aesthetic analysis. Poets and planners, Pooh-Bahs and panjandrums. But land lovers all.

Why isn't this enough, these examples, to show that progress in land ethics is afoot and that eventually all will be well? We have, to be sure, a wilderness policy (although it operates only on the federally owned lands). We have a recreation policy (though emasculated by budget and staff cuts at the federal level). We have a soil erosion policy (albeit weak and pusillanimously voluntary). We have wildlife policies, and historic preservation policies, and we have various state and federal policies to protect places of special significance such as the coastal zone. But, alas, there is no overall policy for *land*.

Leopold insisted on dealing with land whole: the system of soils, waters, animals, plants that make up a community called "the land." But we insist on discriminating. We apply our money and our energy in behalf of protection on a selective basis. Not of land, but "natural areas." Not of land, but "prime farmland." Not of land, but "wilderness." Leopold briefly compares the evolution of the land ethic with the evolution of ethics concerning children in our society. Child labor laws are now applied to all children, and recently the rights of children not to be beaten by their parents have been asserted. Would we say, for example, that we have an ethic for the protection of children in our society but that it pertains only to some children—perhaps those whose noses do not run—and not to others? The question this raises for land and its future is: if an ethic is selectively applied, is it still an ethic? Or is it just a hobby?

The idea of a hierarchy in land quality is, nevertheless, *the* basic tenet of the conservation and environmental movement. We do not see this as an ethical flaw in our thinking, but necessary to the organization of our actions. We preserve prime farmland because it is the most productive in terms of dollars flowing into the agriculture industry. Therefore, we are casual about other lands. Since the 1920s we have been plowing up great swaths of thin prairie soils in the High Plains, soils that should never have seen the "sillion shine" of the plowshare. Now, huge center-pivot irrigation rigs crawl over the land like weirdly articulated steel insects, sucking up the irreplaceable reserves of water in the Ogallala aquifer, the multistate underground lake that when exhausted will leave the land defenseless against the wind that even now piles the sandy soil against the fences, like dunes.

In urban areas, we commonly assume that we must ruin one landscape to preserve another. A beautiful apple orchard becomes uglified by tasteless development because we wish to save a marsh, and we assume, incorrectly I believe, that because all development is ugly, let ugliness reign except in the marsh. The skunk cabbage thrives, but do we? What hierarchical perversity has led to the tawdriness of the so-called "gateway" communities at the entrances to our national parks? How is it that the beautiful village of Taos can be well managed and its historical artifacts protected and yet the road leading to it so blaspheme the surrounding landscape?

But is there no "normative" landscape? Aren't some places better than others, even so? On the one hand, ethologists tell us that all animals have an instinctive habitat preference. *Homo sapiens* is an animal species, therefore humankind has a habitat preference just as imbedded as that in, say, a meadow mouse. Now the anthropologists take over. What is the preferred habitat for humans? Well, look no further than the place where humankind arose, where it emerged as a species. What place is that? The archeologists have the answer: the Great Rift Valley in East Africa. Well, what was it like there a couple of million years ago? It was a savanna, say the paleontologsts. Short grass and scattered trees. Campsites by the lake or best of all where the river runs into the lake. And was that really the good life? Sure, say the prehistorians. A gatherer-hunter spent maybe twenty hours a week gathering and hunting and doing other work necessary to his survival. The land was abundant, the landscape provided safety, the climate was ideal (not as dry then as it is now). So most of the time was taken up with peaceable intellectual, social, and artistic pursuits. Normative landscape? It is, of course, the Garden of Eden, paradise (which is Persian means garden). Where is it today? Why, in the landscaped estates of the wealthy. On golf courses. On wilderness calendars that show alpine meadows rimmed with trees and mountains (as did the mountains rise tens of thousands of feet, to Kilimanjaro, as a backdrop to the Great Rift Valley). And, at the end, the normative landscape is where we are laid to rest. Short grass and scattered trees in manufactured savannas called memorial parks.

On the other hand, other landscapes are good, too. What is better than a bustling, unblighted city neighborhood, a tidy suburb, an apple orchard framing a white farmhouse, a sea-pounded cliff, a deep hemlock forest, an unspurious fishing village, a desert where the cholla and saguaro give spiny protection to a community of strange creatures? Any place is a good place if it is allowed to be true to its inherent nature. Land lovers understand such speculations, understand them to be the eternal argument of culture versus the genes.

Scott Buchanan, the philosopher and educator who originated the "great books" program at St. John's College in 1937, observed that "one of the impressive functions of the cosmic idea is to preside over the birth of possible, new, and good worlds, and to incite new wills to make them actual."

In Leopold's land ethic, we are in the presence of a cosmic idea. It can, and has, incited new wills to make its promise at least partially actual. But I would argue that the great irony of the land ethic is that those who embrace it more fervently—those who love the land—often are among those who obstruct its fulfillment. By and large, land lovers have copped out, arguing for the protection of the land on every socioeconomic basis they can think of in the manner of Smith and Bentham and Locke, save the ethical one. We try to find arguments to suit the cynics: those who, said Oscar Wilde, understand the price of everything and the value of nothing.

As the sun sinks in the West in more ways than one, we search frantically for tiny indications that a land ethic, at least a preclusive one, can really exist in America and that it can exist for reasons of obtaining a good life rather than just making a good living.

The closest anyone has ever come to actualizing what we might call a Level I land ethic has been the state of Oregon, under the leadership of its late governor, Tom McCall. McCall died, of cancer, in January 1983.

He was, in my view, the most effective political operative in behalf of American land since Theodore Roosevelt.

A dozen years ago, Oregon enacted a legislative package concerning land use in the state. It was designed to stop what McCall cheerfully described as the "grasping wastrels of the land," the "buffalo hunters and pelt skinners," those who presided over the "ravenous rampage of suburbia," and infectious "coastal condomania." We would stop cold, he said (and this is my favorite McCallism), "the sagebrush saboteurs." In a nutshell, Oregon's legislation established an independent state-level body that promulgated statewide land use "goals and guidelines" for application via local regulations. If regulations were not applied locally, then the state government would apply them itself. There were nineteen goals-and-guidelines statements dealing with such matters as trasportation, industrial siting, waste treatment, water supply, and farmland. The net was a broad one. Not many square feet of McCall's "beautiful Oregon country" were uncovered by the legislation.

One goal, dealing with agricultural land, is especially instructive. It provides the basis for "exclusive farm use" zoning—EFU—on most existing and potential privately owned farmland in the state. Though the farmland zones are established locally, the permitted uses are defined by state statute, which provides that any new lot in an EFU has to be large enough to maintain a viable agricultural economy in the area so that families can continue to make a whole living from it. If, therefore, farms average 100 acres in, say, a dairying district, then local governments cannot permit the subdivision of land into parcels substantially less than that amount. So where is the new development supposed to go? Why, inside the UGB— the Urban Growth Boundary—established under the legislation to confine urbanization to areas in and around existing settlement rather than let it ooze all over the landscape.

In my view the Oregon story is important not because of the technicalities of its legislation but because in one state at least, a government (which is to say, the citizenry) was able to establish convincingly that "the land" is of *public* concern, not simply a matter that can be left to private economic decision. If it seems like a truism that there are public rights to be considered as well as private ones in the management of the land resource base, please remember that only a very small handful of state and local governments of the United States has any kind of policy dealing resolutely with land in any category, much less comprehensively.

These days we are confronted with a growing trend toward the "privatization" of land-use decision making, to employ an obnoxious contemporary term, which together with a rather negative government role, especially at the national level, seeks to influence land-use decisions so timidly that it is scarcely worth anyone's time messing around with it. We do not seem to be able to produce clear-cut statutory policies that provide, in the law of the land, laws *for* the land. For example, in one lukewarm piece of legislation, recently enacted, the best we can do for the Barrier Islands, those magnificent shifting dunes with their fragile ecosystems that guard our coastline from Virginia to Florida, is to constrain the federal government—the *government,* mind you, that is supposed to be on our side—from not doing anything *itself* to degrade the islands further.

Tom McCall, God bless him, would have none of this pussyfooting around. His approach was Mosaic, with plenty of thou-shalt-nots deriving from a clearly conceived right of the ordinary citizen to have a landscape worth looking at and living in.

But at what a price. In the end, when he was dying, he had, because a referendum had been placed on the ballot to abolish his policies, to convince his fellow citizens once again that their land was precious to them. For his trouble he was told, and not for the first time, that his idea was nothing but thinly veiled Marxism, that what our forefathers fought for, our sacred heritage, was being abridged, that he was depriving his fellow citizens of their constitutional right to destroy the land of Oregon as they saw fit. In 1982, in the midst of this battle, I invited McCall to attend a meeting I had organized in Ohio, and he told me he would come. But later he telephoned; he said that the cancer was kicking up again, taking his strength, and he'd better stay home in Oregon to fight off "the grasping wastrels" once again. He did, and he won: The voters sustained the legislation. But soon after, McCall was gone. We were never to see him again. And his like comes around rarely.

Leaning on champions like McCall reveals a terrible flaw in our perception of how Leopold's cosmic idea is to be made actual, for it is plain to me that the future of the land ethic—if we are to get beyond Level I, even in a single state—cannot rest on the chance that a Tom McCall will meet history in just the right, dramatic way to save the land. And have it stay saved. Let us be honest. There is no real ethic present, no permanent system of values to which we have given our general consent, when the laws expressing it are constantly challenged, vicious arguments mounted, patriotism called into question as it was for McCall himself. The land ethic of the future will not be without its complications in application, but its basic premise must be accepted as being natural and obvious, made manifest simply by "listening to the ground" in the words of the chief of the Western Cayuses. For this, it may be necessary to look not only to the brilliant political apologist and leader, but to the ordinary *users* of the land as well, for they are most in touch with it.

Why should *they* not be the ones to insist on policies for the land's posterity? Indeed, some do, as in the case of Tom Lamm's farmers—though often we do not hear them, so engrossed are we in our hierarchical attitudes about land. But if I were a farmer, I'd rather farm for the future than for a bunch of bankers. If I were a sawyer, I'd be pleased to scratch what Gifford Pinchot wrote into the housing of my chainsaw, that what America *still* needs to understand is that "trees could be cut and the forest preserved at one and the same time." If I were a herdsman, I would want pastures of plenty to the horizon of time as well as space. If I were a fisherman, I would wish for the heavy-bodied salmon to run freely up free rivers forever, squirming into the far pools of ancient memory.

To assume that individual, small-scale use of the land's resources leads inevitably to its destruction is to confuse a failure of policy with the function of land stewardship by those to whom the land has been entrusted for care. In *A God Within,* Rene Dubos has written of his beloved Ile de France country, northwest of Paris, as a landscape that is not only preserved, but improves. He quotes the poet Charles Péguy: *"Deux mille ans de labeur ont fait de cette terre/Un reservoir sans fin pour les âges nouveux."* Dubos translates this: "Two thousand years of human labor have made of this land an inexhaustible source of wealth for the times to come."

Thomas Jefferson had confidence in the American people not only as electors of political leaders but as stewards of land. "I am conscious," he wrote in 1785 to his friend James Madison, president of William and Mary, "that an equal division of property is impractical, but the consequences of this enormous inequality producing so much misery to the bulk of mankind, legislators cannot invent too many devices for subdividing property. . . . The earth is given as common stock for man to labor and live on . . . it is not too soon to provide by every possible means that as few as possible shall be without a little portion of land. The small landowners are the most precious part of the state."

These, the small landowners, and those who would become small landowners, are the people who most validly may insist on morality in land use, in principle as well as policy. It is in their behalf, primarily, that the Bishops have addressed themselves to the land ethic in their pastoral letter.

And yet the future of the land ethic depends no less on the rest of us, as the faithful supporters of stewardship wherever we can find it—in law and in practice. I can't imagine how else it will come about, the actualizing of this cosmic idea, other than in small but vivid increments of individual choice and collective action.

"I will lift up my eyes unto the hills," the Psalmist wrote, "from whence cometh my help." In my former valley, I could still lift my eyes to the San Gabriels, despite the zooming traffic at my feet. The mountains rise abruptly from the settlement below, and when I was a youngster my friends and I would climb the steep folded flank of the first range of a place we called the Lookout. . . .

"Watch out for rattlers," yelled the leader, whom I could still hear across the chasm of time. He was a wiry ten-year-old with carrot-colored hair cut close for summer, which was perhaps the summer of 1940. The clattering stones reminded him of the sandy-brown snakes that liked to hide among the roots and in the crevices to keep cool.

"Rattlers don't worry me," someone shouted back. "It's the mountain lion."

"Mountain lion?"

"Yeah, mountain lion. You never seen a mountain lion?"

"I've seen Mr. Williams' yellow dog. I bet that's your mountain lion."

"It wasn't no yellow dog that chased the coyotes away."

"Shoot."

"Well, shoot yourself."

We pushed ahead, grabbing at ironwood branches to hand ourselves up the slope, winding through dead yucca spires that earlier in the year, in spring, had shot up six feet from the nest of sword-like leaves, each spire surmounted by pannicles of creamy blossoms, like giant lilies, which they were.

Finally, dusty-dirty, with sweat rivers eroding down flushed faces, we gained the Lookout and our California valley was spread out before us.

"Jeez, lookit everything," said the redhead.

"There's the mesa," said another. "See, we're looking *down* on the mesa." It always amazed us, to be above it.

"You mean where the mountain lion scared the coyotes away?"

"Can't you pipe down!"

We knew that if the coyotes were gone from the mesa for an evening or two, they'd soon be back, the moonlight behind them, howling with their muzzles pointed skyward in the yip-yip-yiparoo that covered the sounds of their brothers creeping up on the chicken coops below.

As we looked across our valley, the brightness made it all the closer, like a medieval triptych whose foreshortened perspectives could give a scene a holy quality. The vineyard rows cut into each other at crisp angles, the details of the vine-trunks and the interlacing runners almost visible, even from here. And beneath the gray-green leaves hid the heavy blue and green bunches, like prizes.

"Lookit Mr. Kraus. Lookit!" yelled the redhead again, though only a couple of feet from the farthest ear.

Mr. Kraus, for it was he, was cranking up the tractor, which had a bouncy metal seat, smoothed and glinting with wear. The pops and clanks drifted up to us, out of synch with his actions, for he was at a distance. A tiny speck of red appeared as Mr. Kraus straightened up and ran a bandanna across his flat German brow. He was constructed of planes and bands of bone and muscle. With his sons he would be cutting the grapes with linoleum knives honed like razors. They would pile the dark bunches in wooden crates left at the end of each row, thence to be hefted onto the flatbed wagon, drawn by the tractor.

We watched for a while as Mr. Kraus messed with the spark, then we surveyed other quarters. "The school, the school," someone said, directing our attention to it. And there it lay, its Spanish tile roofs enclosing a yard of live oaks whose cantilevered branches could, when school was in session, support a line of children as a telegraph wire supports a line of swallows.

And so we went through a litany of places, freshly revealed by our superior angle of view: the olive trees along school street, the state highway below, the tomato patch—a huge field of ruby fruit—and the orange groves set into the scene like emerald rectangles spread across the middle distance.

To the northwest was Tujunga, and southeast the Devil's Gate, great arroyos among many that guided the waters out of the mountains during the brief season of rain. At other times, the water for the houses and the groves and vegetable crops (vineyards were dry-farmed) came from catch basins fed by water mines, bored in the canyons by the earliest settlers, perhaps even the rancheros of the original haciendas.

In such canyons we would find smooth stones, and carrying them up the mountain would send them humming aloft from the Lookout, into the brilliant air, with David's-slings made of rawhide thongs knotted to leather pouches that cradled the missiles. We swung them round and round, faster and faster, and when we released them there was only sound, for the stones would fly faster than the eye could follow, upward in a great arc—a fragment of our place flung into a distant land, perhaps another country whose people would marvel at the mysterious object falling at their feet.

From fragments such as these will a land ethic be created, the fragments of a sensibility and a hope whose origins are in the earth itself. The Kodak carousels of our memory go round and round in the darkened living rooms of America, where images of the land are cast upon lenticular screens. "Oooh. Aaah," say the neighbors. "Beautiful. Just beautiful."

And so it is.

22

Environmental Protection: Luxury of the Rich or Requirement of the Poor?

Janet Welsh Brown

IT WAS POPULAR IN THE 1970s for critics of the environmental movement to charge that environmentalism was the "luxury of the rich." The editor of *Harper's* magazine labeled environmentalists "elitists," and at the 1972 United Nations Stockholm Conference on the Environment, representatives of the developing countries charged that environmentalists from the industrialized, rich countries were using environmental concerns to keep the world's wealth to themselves and to prevent economic development of poor countries. Only fifteen years ago it was commonly thought in this country that we would have to choose between jobs and the environment, and that in the Third World, people would have to choose between development and environmental protection.

In the intervening years we have witnessed a remarkable reversal of these beliefs. At home, Congressional Black Caucus members, who tend to be particularly sensitive to issues of poverty, have the best environmental voting record of any group in Congress, because they know that the poor get hurt most by pollution and environmental degradation. And leaders of the developing countries, especially the poorer countries—those with the most fragile lands, or the most problematical rainfall, or the most crowded urban *barrios*—understand full well how future economic development relies on the sustainability of the resource base.

Although the Washington staffs of national environmental organizations are still mostly white, well-educated, and middle class, the communities organizing around the country at pollution sites are not. Blacks and whites organize together against dumping in their communities, against hazards in the workplace, against lead in the air and on the streets. And around the world, people of every color, culture, and political persuasion are organizing—to protect trees in Nicaragua, to protect watersheds and mangrove swamps in the Philippines, to demand dependable water supplies and sanitation in India.

This is the greatest accomplishment of the environmental movement: this revolution in awareness and understanding, this sense of urgency, this knowledge that environmental protection is not the luxury of the rich but a matter of survival of the poor, this realization that we share one, finite earth and that all of us are responsible for what happens to it.

Completing this revolution is also the greatest challenge before the American environmental movement. Every day we learn of new relationships and complexities, and hear the *no easy answers* response. The more we learn, the more difficult it seems to find the right answers.

No, it is not our imagination—things *are* changing *faster* than ever before. In every major sector, the trends that determine the quality of life are accelerating dizzyingly.

TREND ONE: The world's population is growing faster than ever before. Although the rate of increase leveled off in 1972, and declined in many countries, the absolute numbers multiplied because the base is so large. The world's population has doubled since World War II, from 2.5 to 5 billion, and the next billion will be with us by the end of the twentieth century. Nine hundred million more jobs will be needed by then for people already born—some 60 million new jobs every year.

TREND TWO: We are consuming more. Thanks to a strong recovery after World War II and real growth in the majority of developing countries, most of the world's people are better off than they were in 1945, and they command and consume much more than their parents did. Yet, the number of desperately poor also has increased dramatically. A prime indicator is malnutrition: Despite impressive worldwide gains in food production, more people live with hunger and malnutrition in 1980 (before the recent African drought) than did in 1970. Some 800 million eat fewer than 90 percent of the calories deemed necessary by international standards for an active working life. Of these, half eat fewer than 80 percent of the standard—too few calories to prevent stunted growth and serious health risks. Some of the hungry and starving live in the United States or other industrialized countries, but most dwell in the Third World.

TREND THREE: People are producing more waste—human, household, industrial—every year, and the United States consumes and wastes more profligately than any other society. Every man, woman, and child produces nearly a ton of trash per year. While we make modest gains in reducing industrial pollutants and recycling community waste, we are becoming more and more of a throw-away society. Increasingly, hospitals and other large institutions discard rather than reuse, and even more of our meals are "carry-out," packaged and served in disposable containers. We Americans pay as much for packaging our food as the farmers make for growing it.

TREND FOUR: The earth itself is beginning to be overwhelmed by human activity. The accumulated warming effect of carbon dioxide and other greenhouse gases is becoming evident. Ozone is disappearing much faster than scientists thought even a year ago. Deforestation and desertification have picked up speed, as has the accompanying loss of species. Hydroelectric dams silt up faster than imagined. More and more groundwater is polluted. Irrigated lands become waterlogged and salinated as fast as new acreage comes under irrigation. Soil loss is rapid, in India, Africa, Guatemala, and the United States.

TREND FIVE: There are more refugees than ever before in history, many of them fleeing poverty and environmental degradation. As many as five million Ethiopians have been displaced from their barren, eroded highlands, half a million of them across the border to neighboring countries. Since 1980 at least a million Javanese have left their resource-battered, overpopulated island for Indonesia's outer islands. Deforestation, soil erosion, and water depletion, intensified by a rapacious dictatorship, created the tragic spectacle of the Haitian boat people, only some of whom made it to Florida's shores. For every two Mexicans who leave their impoverished countryside for Mexico's cities, one crosses the border to the United States.

TREND SIX: The nations of the world are experiencing rapid, sweeping changes in their financial fortunes. Higher and higher portions of national budgets are spent on arms, now $1 trillion a year. And international debt is at an all-time high, with the developing world's debt of the early 1980s now exceeded by that of the United States. In 1985, the dollar reached its highest value in history, then steadily declined. The stock market reached its highest point ever in 1987 and then came crashing down further and faster than in the Great Depression, dropping 508 points in one day.

TREND SEVEN: Technological change is proceeding equally quickly. Communications and transportation breakthroughs translate into higher mobility, enabling the AIDS virus, emergency signals, and the latest Tokyo stock prices to travel very fast. Advances in genetics and biotechnology open vast new possibilities in agriculture and health—and possible dangers as well. They could create nitrogen-fixing or drought-resistant crops to restore eroded lands in semiarid Africa and a vaccine to wipe out malaria's debilitating effects—or, conversely, release into the environment some highly resistant pest that destroys food supplies. The breakneck development of materials technology such as superconductors has left even scientists breathless. The human ability to detect the smallest units of life and energy, to measure minute amounts of toxins, and to monitor the weather and forest loss outstrips society's ability to use the information.

The acceleration of these and many other trends sets our era apart from all of human history. So many interactions, so fast, make us positive of only one prediction: more change, more surprises, less predictability.

Two imperatives arise out of these supercharged trends. They are the twin messages of *Our Common Future,* the 1987 report of the World Commission on Environment and Development: that the well-to-do industrialized nations must learn rapidly to do more with less, and that poor developing countries must tackle poverty and environmental degradation together. Neither of these imperatives fits easily with our national obstacles-be-damned character, and both force American environmentalists to deal with issues outside their traditional realm. Herein lies the challenge to the environmental movement in the next decade.

The United States has long had the world's most productive economy. Drawing on our own seemingly limitless resources and augmenting them with imported foods and materials and manufactures, we have built an enviably high standard of living. Although the habit of fouling our nest and the existence of an underclass testify to an imperfect and sometimes unjust system, the United States has become the model for much of the rest of the world. The American industrial machine, our material values, our technologies, our consumption, and our popular culture are admired and aped all over the world.

Since Earth Day in 1970, the rejuvenated American environmental movement also has inspired Western Europeans, then Third World citizens, and finally activists in the Soviet bloc to galvanize the public and their governments to protect the environment. In Malaysia Friends of the Earth (FOE) literally went to the barricades and to jail in 1987 in defense of their remaining rainforests. In October 1988 the Polish FOE affiliate, *Polski Klub Ekologiczny,* hosted in Krakow the first meeting of East and West European environmental organizations. Concern for the resource base is now shared worldwide, and the U.S. movement deserves much of the credit.

There is the dilemma: We in the United States cannot go on consuming at current rates and inviting everyone else to follow our example unless we do so much more efficiently. We need incentives for wise resource use and stiffer punishments for abuse or indifference. We need better technology and a return to such old American values as Yankee ingenuity and Ben Franklin's "waste not, want not."

Unfortunately, these challenges come at a time when Americans are riddled with self-doubt and most American leaders lack courage and imagination. We have lost some of our faith in technology, and we are frightened by the market crash and worried about the steep climb in trade deficits and debt. We sense that U.S. hegemony is declining, that the end of the American empire may be in sight. Never mind that it was not ever complete hegemony or a real empire—the shifts in fortune and power still make us uneasy. Americans do not greet trouble with Calvinist discipline and self-denial. If the pie is too small, Americans would prefer not to divide it differently, but to make a bigger pie instead. Our seemingly infinite resources and energy, coupled with our faith in technology, make us bridle at limitations. And our political leaders know that, outside of war-time, moralistic calls for belt tightening do not win elections. Therefore, we blame Japan for our trade problems, raise barriers to textiles from Asia, and require our contractors and clients to "Buy America." In general, we react to international and global problems with a mind set that is still nationalistic.

The nation needs new political leadership that is inspired by the environmental lessons of the recent past. The environmental movement has served the nation well in the last decade and a half, and can serve the world well through the end of this century.

What have we learned? We learned that pollution and degradation are expensive and that it costs more to repair and restore than to prevent in the first place. We learned that the polluter should pay. We learned that the price of industrial and household waste disposal is high and getting higher, that recycling, reusing, and reducing pollution at the source could pay off. We learned, in one industry after another, that regulations can stimulate the development of cleaner and more efficient technologies. As we began to understand persistence and resistance, we learned something of the ecological and health costs of excessive chemical pesticides use. We learned to make more efficient appliances (and finally to agree on standards for them) and to build vehicles that get better fuel economy. In our western states we learned that subsidized irrigation prices could lead to costly environmental practices like waterlogging and salinization. We learned that electric rates could affect energy efficiency positively or adversely. We learned the cost of soil loss brought on by poor farming practices and by incentives to farm marginal lands.

In developing countries, we learned the value of rainforests and the cost of soil loss—not only to the farmer, but also to mangroves and spawning grounds. We began to understand that the lending practices and development policies promoted by the multilateral banks and our own Agency for International Development sometimes had destructive side effects that made them both economically and environmentally unsustainable. We learned that, especially in the Third World, poverty and environmental degradation go hand in hand, and that one cannot be stopped without stopping the other. We learned that in both the industrial and developing worlds we can increase production while using less energy, through increased efficiency and conservation. We learned that nuclear war would be the ultimate and final environmental insult and that the disposal of nuclear wastes had to be solved before nuclear power could be justified.

Environmentalists, their organizations, and their movement deserve credit for this progress. The Environmental Defense Fund, the Natural Resources Defense Council, Friends of the Earth, the Audubon Society, the Environmental Policy Institute, the Rainforest Action Network, the National Coalition Against Pesticide Misuse, and all the other networks and coalitions made this understanding possible. Their insights, their persistence, their legislative acumen, legal prowess, scientific analysis, and media proficiency brought Americans to that understanding.

In short, we learned that environmental issues are embedded in economic ones and that no economic strategy is without environmental consequences. That lesson, though still imperfectly internalized in decision making in our own country, will see us safely through this century and into the next. It also provides the basis on which Third World development must proceed.

The tasks are difficult and the job has just begun. The environmental movement now has to stretch and apply the lessons it has learned in even more demanding, largely international ways. The movement must learn more about the people and cultures of the Third World, their capabilities and leadership, and their development needs. It must master the tough issues of U.S. trade policy, debt management and relief, and international finance. It must recognize the unyielding institutional barriers erected by our national addiction to consumption over savings, the effect of concentrated corporate power, and the drain of vast military budgets. And it must educate the public and politicians to all of these. For saving the planet means delving into all of these issues and fathoming the relationships among them.

What does this shift in political consciousness require of U.S. leadership? Certainly, it requires policies that maximize the efficient use of resources and curb pollution at home. It requires that we save existing green spaces and wetlands and plan orderly development that preserves places of natural beauty—not only those grand and rare and exotic national monuments and wonders but also the hills around a city, the small urban parks and river banks, and the city harbors where ordinary people can relax as well.

The job ahead requires a massive education campaign about the economic and aesthetic value of ecological balance and about the human condition and development needs of poor people in the Third World. It requires the public will and the policies and programs needed to eradicate Third World poverty.

The challenge requires a new kind of U.S. leadership in dealing with Japan, the Soviets, the European Economic Community (EEC), and China— the next decade's big actors on the international scene—for we cannot save the planet by ourselves. If the EEC continues to dump subsidized agricultural products on world markets, if Japan's markets remain closed to Third World manufacturers, if China and the Eastern bloc countries continue unmitigated carbon dioxide emissions, if the Third World debt goes unattended by others, then even the most enlightened U.S. policies could be neutralized. A global understanding of the world's environmental problems will necessarily lead to a sharing of the burdens of achieving sustainable development. The environmental movement's contributions must include steady pressure on our own government and, through our environmental colleagues in other countries, on foreign governments to reorient their domestic and international policies to sustainable development. Going it alone will not work; the global nature of the problems requires multilateral approaches.

It is time for a new American hegemony based on moral, humane, and technical strengths, on a stewardship of earth and its people that takes advantage of American idiosyncrasies—our optimism, our generosity, our love of nature, our growing awareness of environmental threats, our strong independent sector, and our technological ingenuity.

23

Forging a Viable Future

Richard S. Booth

ABOUT A DECADE AGO, the year 2000 appeared to provide a comfortable time frame for undertaking what then seemed to be far sighted planning regarding use of the world's resources. As that date draws closer and that planning remains largely undone, three critical points are clear. First, we are inflicting tremendous and constantly increasing damage on all components of the environment, including ourselves. Second, we are engaged in a race against time to change our actions before the natural web that sustains human society unravels to an irreversible degree. Third, we are losing significant ground in that race.

Many authors have examined the basic causes of the environmental crisis. They have pointed, for example, to the evolution of our political and economic institutions, to our religious traditions, to our increasing separation from the land, and to our growing reliance on ever-expanding technologies. In the end, our most sophisticated analyses of the causes of the environmental crisis appear to come down to something very basic: namely, as so-called modern society experienced several centuries of continually expanding power to manipulate the environment, it increasingly ignored the fact that its welfare depends on the proper functioning of the natural world.

During most of that history, we did not seriously threaten the entire global environment because the earth was so large compared to the scale of past human activities and because the environment is inherently resilient. Those circumstances, however, are now changing rapidly. Today, we significantly affect the entire biosphere, and we are taxing the environment's resilience beyond its limits. Our abuse of the environment is inflicting grievous harm on the natural world's ability to function properly, and we will pay increasingly for that abuse.

Evidence of the seriousness and pervasiveness of the environmental crisis surrounds us. First, consider the human-environment relationship as a contest in which points are won when we balance our actions with the needs of the natural world and lost when we fail to do so. We seem determined to lose this contest by a very large margin. For every threatened species we protect (points won), we threaten or extinguish many others because species disappear before we even know they exist (points lost). For every piece of land we use properly, and for every piece of fragile land we preserve, we consume and abuse countless others. For every pollution source we clean up, we fail to prevent or even discover many others. In spite of determined efforts to protect the environment, we are losing far more points in this contest than we are winning. The points won at best represent largely separate and scattered successes. These successes prevent or ameliorate some environmental abuses, but by no means all of them and usually not even the worst ones.

Second, consider the major problems now confronting us. A realistic analysis of those problems supports the conclusion that environmental abuse directly causes many of these problems and contributes significantly to many others. Environment-related problems plaguing different parts of the world include, among many others, starvation in sub-Saharan Africa, the proliferation of chemically contaminated water supplies across the United States, increasing human health problems caused by environmental contamination, the serious deterioration of productivity of the world's agricultural lands, and particular disasters such as Love Canal, Bhopal, and Chernobyl. Environmental abuse problems also encompass the potential for international conflict over the availability of oil and natural gas supplies (or other vital natural resources) and the impoverishment of millions of people whose livelihoods depend on the continued use of natural resources (such as forests) that are being seriously degraded. From a still wider perspective, environmental abuse is tied inextricably to the enormous inequalities that characterize our world, to the maintenance of repressive governments in many nations, and to economic arrangements that have placed many Third World nations so deeply in debt.

Third, consider the dominant characteristics of modern society. We are generating greater and greater conflicts between human beings and the natural environment. For example, the products of modern technology multiply rapidly, even though many of these products cannot be used for their designed purposes or disposed of without causing serious environmental disruption. The volumes of waste products produced by those technologies increase at alarming rates. The phenomenon of use-and-throw-away products now dominates many of the developed world's major industries, and developing nations seem determined to follow the same course. Environmental manipulation with pesticides and other chemical materials has become an addiction, a habit we seem unable to break even though it increasingly threatens our well-being. Although the rise in population growth rates has eased recently, the world's human population continues to climb steeply, imposing greater and greater burdens on the earth's resource base and condemning millions to lives of overwhelming poverty. The Third World is experiencing the phenomenal growth of enormous urban centers with complex environmental problems that seriously threaten the lives and safety of large percentages of those cities' populations. Notwithstanding some advances on nuclear arms control, the world's superpowers go on expanding their nuclear, chemical, and biological warfare technologies, even though the use of just a small portion of those arsenals would largely destroy the world as we know it.

Fourth, consider the implications of our rapidly expanding scientific understanding of the environment. Through increasingly sophisticated technologies we are gathering more and more evidence about the causes and effects of acid rain, destruction of the world's renewable resources, human impact on the global climate, destruction of the ozone layer, and the pervasiveness of manmade contaminants throughout the biosphere. In its best light this evidence documents the global implications of what we are doing to the environment. In a more somber light it may document the beginning unraveling of the environmental web that sustains all life on this planet.

In spite of the multiple factors that underscore the seriousness and pervasiveness of the global environmental crisis, we observe a deeply troubling phenomenon: namely, that neither the people of the world nor their leaders are taking decisive action to confront it. Although many environmental protection efforts do exist, in both public and private sectors and in many countries, these efforts are scattered and piecemeal, and they are doing relatively little to alter either the overall pattern or impact of our assault on the natural world.

We should assess very carefully our inability to confront the environmental crisis systematically. Consider these facts. Few of our graduate and professional schools have begun to address seriously the critical social changes needed to resolve the environmental crisis. The global environmental crisis is almost a nonissue in American politics. The leaders of the world's two most powerful nations can meet and discuss world issues without giving any serious attention to the plight of the global environment. No major government leader of any country is making a critical priority for his or her government the need for fundamental changes in how society affects the environment.

The necessity of balancing human actions and environmental needs remains outside the central factors that guide our modern world. Of course, we sometimes recognize the environmental implications of our actions, but usually we fail to do so. Of course, political leaders sometimes consider environmental factors in making decisions, but when they do, they almost always relegate those factors to secondary, nonessential levels of importance. Of course, environmental considerations shape certain choices, but almost never does the environment play as major a role in social choices as, say, economics, security, national prestige, or vested interests.

In short, neither the world's citizens nor its leaders are doing a great deal to wrestle with this global crisis. When we ask why this is so, we can identify numerous factors that contribute significantly to the neglect and passivity with which most of us confront the environmental crisis. Although this listing is not comprehensive, the following factors account for much of our collective neglect of this peril.

To a very large degree we do not recognize the direct relationship between harm to the environment and harm to people. We continually fail to see that hurting the environment hurts human beings in countless ways, and we pay for that failure repeatedly.

People often ask whether this nation (or any other) can afford to restore its abused farmland, or clean up its abandoned chemical waste dumps, or prevent contamination from nuclear power plants, or protect its coastlines and riverbanks from despoliation, or cleanse the emissions from coal-fired power plants. These questions miss the most critical point. If a nation fails to pay the costs of avoiding or repairing environmental damage, that damage will extract even greater costs in the future—costs measured in terms of weaker bodies, higher levels of disease, shorter lives, greater numbers of stillborn babies, more contaminated homes, and many other indices of human misery. The central issue raised by environmental abuse is never whether we can afford to prevent it or to clean it up once it has occurred. Rather the issue is simply who will pay for that abuse, when and how will it be paid for, and how large the cost will be.

Our failure to recognize the link between environmental abuse and human welfare stems from our lack of environmental education. In spite of our increasing scientific sophistication, most of us understand very little about how our actions are upsetting the essential balance of nature. We believe we can take whatever we want from the environment, no matter how much, and dispose of whatever we want to, no matter how dangerous. Simply put, our greatest challenge in meeting the environmental crisis may lie in the tremendous task of educating people about the natural world.

Most of us fail to recognize that environmental abuse can lead to worldwide catastrophe. We believe that war and economic depression present the greatest threats for global disaster, and certainly the ravages of world war and depression in this century provide some basis for those beliefs. Most of us do not understand that Armageddon is as likely to arrive in the form of an environmental disaster as a nuclear holocaust.

Our inability to recognize the possibility of a global environmental disaster arises to a significant degree from our collective misunderstanding of time, history, and human nature. Somehow we believe that because the environment has tolerated the current level of exploitation in the very recent past, it can and will go on doing so. We fail to appreciate, among other things, that past societies, such as that of Babylonia along the Tigris and the Euphrates rivers, disappeared because they did not live in balance with the natural world. We fail to appreciate that in recent decades we have dramatically escalated our assault on the environment. We fail to appreciate that the cumulative effects of environmental abuse may take years to reach the moment of supreme crisis and that once the critical moment arrives, natural systems may unravel with dizzying speed. We fail to appreciate that we are not likely to be able to act wisely or quickly once that moment of absolute crisis arrives and that the way to avoid the consequences of that crisis is to act now to design and implement collective solutions.

This failure to foresee the possibility of global disaster stems also from the fact that most of us do not see that two radically different circumstances are feeding the environmental crisis. First, the world's developed societies abuse the environment, because they use far more resources than they reasonably need and impose staggering amounts of waste products on the natural world. Second, many of the world's undeveloped societies abuse the environment because their people are desperately trying to scrape together enough of life's essentials—e.g., food, water, shelter, and heat. In short, wealthy people abuse the environment, because they use and throw away too much, and poor people abuse the environment, because they have too little. Wealthy nations suffer from industrial pollution, oil scarcity, and destruction of wildlife and wilderness. Poor nations suffer from famine, abject poverty, and disease. Both sets of problems are driving us toward disaster.

Neither of the dominant economic paradigms that guide human interactions—namely, capitalism and socialism (or whatever other labels we assign these two major sets of economic thought)—constructively addresses the necessity of our living in balance with the natural environment. Both paradigms encourage us to exploit and despoil the environment for short-term gains. As a result, our most fundamental arrangements for conducting human interactions largely ignore the human-environment relationship.

Regardless of our preferences for one or the other of the world's dominant economic systems, we should recognize that as far as the environment is concerned, both are leading us toward disaster. Capitalist societies are doing their utmost to exploit the environment as rapidly as possible in order to improve the short-term well-being of their citizens. Perhaps motivated by different impulses, most if not all of the socialist societies of the world are doing the same thing.

Sovereign nations play the most basic and important political roles in the international arena. These dominant political actors do not recognize the critical relationship between their well-being and the quality of the environment. They are unwilling to set aside their parochial differences and address together the major social, economic, and political factors that are rapidly expanding the environmental crisis. Their values and their deeds impede the joint, long-term, and mutually beneficial actions necessary to resolve this crisis.

The self-centered and selfish roles played by sovereign nations reflect the essential character of international relations. We have long lived in separate nations; we do so now and perhaps always will. Because of our separations, every nation nearly always thinks first of its own benefits, and each acts to serve its own interests, rather than those of the larger community. Existing alliances for military or economic purposes among a few nations do not change this reality. Only in matters involving the nonessentials of international discourse do nations truly work together for the common welfare of all.

We do not appear nearly ready to change the existing pattern, whether for the purpose of feeding the world, preventing its destruction by war, or avoiding environmental catastrophe. Harsh economic competition between and among nations for the resources and markets of the world stands as the basic model of global interaction. That competition flourishes, even though it condemns millions of people to lives of grinding poverty and awards others far more of the earth's bounty than any rational allocation of its resources would allow. Armed conflict remains a commonly used mechanism for settling differences between and among nations, notwithstanding the enormous destructiveness of modern warfare.

Most of us do not see that many of the same forces lead to abuse of both the environment and human rights. Ignorance, selfishness, hate, cruelty, greed, a desire to dominate, and a willingness to harm the weak cause people to violate the land, indiscriminately destroy other species, and pollute the biosphere. These same forces lead to poverty, intolerance, racism, discrimination, and numerous other forms of injustice.

Today, the struggle for protecting basic human rights remains one of our preeminent challenges. Millions of human beings live without the essential ingredients of human dignity: namely, freedom to believe, worship, speak, and associate with others; freedom to work, grow, and learn as each individual chooses, so long as others are not harmed; freedom to live without fear of discrimination and intolerance; freedom to choose and live under a government that derives its authority from the consent of the governed; freedom to live without poverty and to enjoy the fruits of one's labors; and freedom to live securely in peace.

Environmental abuse destroys not only the natural world but also the social, economic, and political conditions that are prerequisites to the survival of human rights. Those rights require tolerance, forbearance, common decency, and respect for one's fellows so that all of us can live with dignity and security. Dignity and security cannot survive long in a starving, grasping, desperate world where a barren and unsustaining environment encourages and forces people to seize the bare necessities of life by taking the possessions, the rights, and even the lives of those who are less strong.

Conversely, human rights abuses destroy not only the lives of individuals but also the levels of human tolerance, sensitivity, and forbearance that are prerequisites for the survival of a high-quality, self-sustaining natural world. A quality environment cannot survive over the long term if we do not treat it with care and respect. Care and respect for nature are not likely to survive in a world willing to tolerate widespread abuse and destruction of the most basic rights of human beings.

The environmental crisis threatens global disaster, and to date we have done relatively little to avoid calamity. To meet the challenge of the future we must do the following: 1) fight environmental abuse wherever it occurs in order to reduce the damage we are inflicting on the environment and to buy time for pursuing the other four approaches discussed here; 2) educate citizens and leaders about the relationship between human welfare and environmental quality, the causes of the environmental crisis, and steps needed to resolve this crisis; 3) redefine key concepts, including critical economic and political principles, so that our basic interactions can be reoriented toward achieving a sustainable balance between human society and the natural environment; 4) expand existing institutions and build new ones designed to sustain and nurture human activities that occur in balance with the natural world; and 5) join environmental protection efforts and human rights protection efforts as indivisible parts of a unified strategy for forging a viable world.

Fighting Environmental Abuse. The major goal of this first approach must be to purchase the time required to effect the four other approaches. By reducing the level of abuse we inflict on the environment, it is hoped that we can delay the unraveling of the environmental web until other measures designed to achieve a sustainable balance with the natural world can make a substantial difference.

The arenas for these holding actions will range from legislatures, corporate board rooms, and meeting rooms to logging sites, garbage incinerators, and hazardous waste dumps. They must proceed in the world's largest cities and its most remote wilderness areas. These actions will require an equally broad range of methods and approaches, including legislation, litigation, boycotts, demonstrations, protests, electoral politics, labor initiatives, and media events.

This approach enjoys the advantage of being familiar. Some people have been fighting these holding actions for a long time. Let us hope that they will continue. Eventual success in dealing with the environmental crisis requires that we increase dramatically the number, scale, and intensity of these actions. In order to buy the time needed, these existing efforts to fight environmental abuse will have to double and redouble until they begin to change the entire course of world events.

Educating the Public. Public education about the environment forms the second broad approach for overcoming this crisis. Environmental education efforts must dramatically increase our collective understanding of the dynamic, essential, and mutually supporting interrelationships that characterize the natural world, including our own role in that world. Further, they must stress how much we depend on a properly functioning

303

natural world, the causes of the environmental crisis, and the means by which we can overcome it.

Only through education can we hope to alter fundamentally widespread patterns of environmental abuse. If we cannot change our thought patterns regarding our relationship with the natural world, there simply are too many ways that serious environmental abuse will continue and increase. If balancing our actions and environmental needs does not become a basic characteristic of all important choices and decisions, no laws, regulations, government programs, market incentives, or any other mechanisms are going to change the disastrous course on which we now find ourselves.

Environmental education on the scale needed will present daunting challenges. Strong and imaginative environmental education efforts will have to proceed at all levels—from the youngest of us to the oldest and from our most practical and straightforward actions and decisions to our most conceptual and theoretical. Both the public and private sectors will have to play significant roles in environmental education, and we will have to ensure that this education permeates our values and our choices.

Redefining Key Concepts. Redefining key concepts that are taken for granted by most segments of modern human society forms the third broad approach that we should pursue. Many of our most fundamentally critical concepts for organizing our understanding of the world ignore the necessity of balancing what we impose on and extract from the natural world against what the environment can in fact sustain. To a very significant degree environmental protection efforts do not require more money or other resources nearly as much as they require new ways of thinking about economic strategies, defense priorities, health and welfare policies, international relations, and so many other aspects of human decision making.

In the United States, for example, we will have to alter our basic notions about property rights that allow a landowner to take actions that will harm the general welfare significantly. We will also have to reconsider our basic notions about public/private relationships that prevent government from determining what products and byproducts industry may produce. We certainly will have to alter our common belief that every person's highest calling is to acquire the greatest possible financial gain.

Among the most important targets for redefinition must be society's fundamentally critical concepts about economics and about the role of sovereign states. Because of those concepts, our major economic and political systems are not functioning in ways that will bring about a balance between human needs and environmental constraints. We must find ways to redefine the theoretical frameworks that govern our major economic systems, regardless of economic ideologies, and that govern the roles played by the world's nations, regardless of political ideologies.

We cannot afford much longer to decide how we will utilize the earth's resources based on economic concepts that discount the value of any natural resource to zero in less than fifty years after it is consumed or lost. We can no longer go on ignoring the possibility that the natural resources we consume or contaminate today will not be available in the future. The world requires a conceptual economic framework that will assess human use of the earth's resources in terms of the sustainability of our actions in the foreseeable future, say five or seven generations from now. Careful planning and resource conservation and recycling must become the norm, not the exception, in dealing with the natural world. We cannot afford much longer the practice of sovereign nations of defining their primary goals as achieving military and economic superiority. We need a conceptual political framework that allows nations to understand that over the long term their survival, security, and well-being are tied inextricably to the ability of every country to feed, clothe, house, and care for its population, all of which depend on a properly functioning environment, and not to acquiring military armaments or to competing with other nations for the world's resources.

Thinkers ranging from Wendell Berry to Lester Brown to Amory Lovins are trying to redefine economic and political concepts that guide us toward a sustainable future. Their efforts, joined by those of others, will have to expand dramatically and relatively quickly. Theoretical and applied ideas in many forms will have to challenge and eventually overcome well-entrenched, and in some cases centuries-old, ideas about economics and about the roles of nation-states. Success will not come easily. This should not surprise us, however, if we recall how well and for how long most of human society has ignored or resisted the powerful ideas of Henry David Thoreau, George Perkins Marsh, Rachel Carson, and others who pointed the way toward a viable future.

Developing Environmentally Sensitive Institutions. Building institutions that will achieve and nurture a rational balance between human actions and environmental needs forms the fourth broad avenue we need to follow. We generally deal with large issues through a wide variety of institutions—e.g., social, economic, religious, legal, and political. How our institutions react to the environmental crisis will determine to a large degree whether we confront it successfully. Just as we will need environmentally sensitive people to win this struggle, we will also need institutions that will protect the environment rather than exploit it.

We must develop new, environmentally sensitive institutions and greatly expand existing institutions that exhibit desired characteristics. These new and modified institutions will have to exhibit numerous characteristics not usually found in existing ones. For example, they will have to demonstrate high degrees of technical proficiency and address problems in a holistic context that encompasses the relevant social, economic, and environmental concerns. They will have to recognize how many different strands of human action intersect and affect the multiple cycles of the natural world. They will have to understand that solutions to problems must both meet human needs for decent lives and allow natural systems to function properly over the short and long terms. Our future institutions will have to make choices that serve broad social needs while never surrendering to narrow parochial interests and never losing sight of the impacts those decisions may have on particular societal elements. They will have to develop incentive mechanisms that encourage people to respect the environment's needs, help ease the necessary transition from today's world to a sustainable future, and find ways to involve affected citizens in their decision making. These institutions usually will have to make difficult decisions in the face of competing priorities and with incomplete information, and they will have to retain the flexibility necessary to alter their strategies and tactics as new and better information about the environmental crisis makes course changes necessary. Most significant, they will have to think about global issues while taking steps to address local problems, and to consider local issues while making decisions about global concerns.

We will need a variety of institutions that in combination can provide the necessary strengths and skills. Fortunately, we have already created some that reflect many of the needed characteristics. We can find useful models, for example, in various consumer cooperatives, many citizen action groups and nonprofit entities, various private businesses and educational institutions, numerous religious organizations, and even some government agencies and international bodies. To date, these models for future action reflect only a tiny percentage of the environmentally sensitive institutions we will need, but they do offer hope that making the necessary institutional changes lies within our capabilities.

Joining Environmental Protection and Human Rights Protection Efforts. The fifth broad approach for successfully confronting the environmental crisis involves joining environmental and human rights protection efforts into an inseparable union. We need to join these efforts in the broadest sense—in articulating philosophy, identifying goals, establishing priorities and agendas, choosing strategies, and taking action. The combined forces of these now largely separate efforts can forge a viable future.

Environmental protection and human rights protection form prerequisites for each other on this crowded planet. Together they constitute the essential building blocks of a sustainable future for the entire biosphere. We can either save both or lose both; we cannot save one without the other. If we do not protect the natural world, efforts to protect human rights must fail as the environment becomes less and less capable of supporting human life and billions of people become more and more desperate and violent in their struggle to survive. If we do not protect human rights, efforts to protect the environment must fail as human society loses its capacity for equity, tolerance, forebearance, wonder, and grace and as people become increasingly harsh and insensitive in dealing with each other and, as a result, with the natural world around them. If we do not protect both the environment and human rights, a deteriorating environment will provide the rationale for abusing human rights; and, conversely, deteriorating conditions for human beings will provide the basis for further abusing the environment.

The environmental crisis presents at least as large a question regarding what type of human beings we are and will be as it does regarding whether we will destroy the physical capacity of the natural world to sustain us. We know that some of us could survive for a considerable time, perhaps indefinitely, on an earth with all other large mammals extinguished; with beauty destroyed; with lands stripped of their fertility and their abundance of plant species; and with grossly polluted air, fresh water, and oceans. We know that in order to do so, those who wish to survive would have to take what they needed from those less able to protect themselves, without worrying about the human consequences of those actions. We know also that billions of people could survive physically, at least for a considerable time, in a world devoid of human rights, but that world would be a truly horrible one.

The environmental crisis presents enormous difficulties. In spite of its pervasive presence, most of human society refuses to take it seriously or to begin to take the steps necessary to confront it successfully. Numerous factors contribute to our lack of attention to these dangers, and environmentalists and others who would try to change our attitude toward the environmental crisis must address and overcome those factors.

Eventually resolving this crisis successfully will require bold actions in what are now largely unchartered territories. The five broad courses of action addressed here should form the core of our efforts. Each presents great challenges. Collectively they challenge us to climb to greater heights than we have ever had to scale.

No one said that resolving the environmental crisis would be easy. It cannot be. It will not be. In his now classic book *The Closing Circle*, Barry Commoner stated: "Anyone who proposes to cure the environment crisis undertakes thereby to change the course of history." We must get on with that task.

24

The Ecological Age

Thomas Berry

AS WE THINK OUR WAY through the difficulties of this late twentieth century, we find ourselves pondering the role of the human within the life systems of the earth. Sometimes we appear as the peril of the planet, if not its tragic fate. Through human presence the forests of the earth are destroyed. Fertile soils become toxic and then wash away in the rain or blow away in the wind. Mountains of human-derived waste grow ever higher. Wetlands are filled in. Each year approximately ten thousand species disappear forever. Even the ozone layer above the earth is depleted. Such disturbance in the natural world coexists with all those ethnic, political, and religious tensions that pervade the human realm. Endemic poverty is pervasive in the Third World, while in the industrial world people drown in their own consumption patterns. Population increase threatens all efforts at improvement.

Such a description of our human presence on the earth tends to become paralyzing. While that is not my intention, it is my intention to fix our minds on the magnitude of the task before us. This task concerns every member of the human community, no matter what the occupation, continent, ethnic group, or age. It is a task from which no one is absolved and with which no one is ultimately more concerned than anyone else. Here we meet as absolute equals to face our ultimate tasks as human beings within the life systems of the planet Earth. We have before us the question not simply of physical survival, but of survival in a human mode of being, survival and development into intelligent, affectionate, imaginative persons thoroughly enjoying the universe around us, living in profound communion with one another and with some significant capacities to express ourselves in our literature and creative arts. It is a question of interior richness within our own personalities, of shared understanding with others, and of a concern that reaches out to all the living and nonliving beings of the earth, and in some manner out to the distant stars in the heavens.

This description of personal grandeur may seem an exaggeration, a romantic view of human possibilities. Yet this is the basis on which the human venture has been sustained from its very beginning! Our difficulty is that we are just emerging from a technological entrancement. During this period the human mind has been placed within the narrowest confines it has experienced since consciousness emerged from its Paleolithic phase. Even the most primitive tribes have a larger vision of the universe, of our place and functioning within it, a vision that extends to celestial regions of space and to interior depths of the human in a manner far exceeding the parameters of our own world of technological confinement.

It is not surprising, then, that when a more expansive vision of the human breaks upon us at this time it should come as a shock, as something unreal, insubstantial, unattainable. Yet this is precisely what is happening. The excessive analytical phase of science is over. A countermovement toward integration and interior subjective processes is taking place within a more comprehensive vision of the entire universe. We see ourselves now not as Olympian observers against an objective world, but as a functional expression of that very world itself.

What has fascinated the scientist is a visionary experience that is only now coming to conscious awareness in the scientific mind. It can hardly be repeated often enough that the driving force of the scientific effort is nonscientific, just as the driving force of the technological endeavor is nontechnological. In both instances, a far-reaching transforming vision is sought that is not far from the spiritual vision sought by the ancient tribal cultures, as well as by the great traditional civilizations of the past. Only such a visionary quest could have sustained the efforts made these past two centuries in both science and technology. Nor could anything else than entrancement have so obscured for scientists and technicians the devastating impasse into which they have been leading the human venture.

Not until Rachel Carson shocked the world, in the 1960s, with her presentation of the disasters impending in the immediate future was there any thorough alarm at the consequences of this entrancement. That is what needs to be explained—our entrancement with an industrially driven consumer society. Until we have explained this situation to ourselves, we will never break the spell that has seized us. We will continue to be subject to this fatal attraction.

To bring about the closing down of the life systems of the planet on such an order of magnitude is obviously not something that originated yesterday or something that arises out of some trivial miscalculation, academic error, or ideology such as the Enlightenment, or even out of the industrial age itself. These are all symptoms and consequences of a vast turn in human consciousness that originated deep in the origins of the human process itself. It must indeed have been associated with those revelatory experiences that we consider the most profound experiences ever to take place in human intelligence, experiences so profound that we consider them to come from some divine reality. Our deepest convictions arise in this contact of the human with some ultimate mystery whence the universe itself is derived.

Some insight may be gained into these issues if we consider that in our early tribal period we lived in a world dominated by psychic power symbols whereby life was guided toward communion with our total human and transhuman environment. We felt ourselves sustained by a cosmic presence that went beyond the surface reality of the surrounding natural world. The human sense of an all-pervasive, numinous, or sacred power gave to life a deep security. It enabled us over a long period of time to establish ourselves within a realm of consciousness of high spiritual, social, and artistic development. This was the period when the divinities were born in human consciousness as expressions of those profound spiritual orientations that emerged from the earth process into our unconscious depths, then as symbols into our conscious mind, and finally into visible expression. All that we have done since then has taken the same course. The divinities have been changed, the visible expression has been altered, but the ultimate source of power still remains hidden in the dynamics of the earth and in the obscure archetypal determinations in the unconscious depths of the human mind.

These primordial determinations were further developed in the age of the great traditional civilizations of the Eurasian and American worlds, the age of Confucian China, of Hindu India, of Buddhist Asia, of the ancient Near East, of Islam, of medieval Europe, of the Toltec, Mayan, Aztec, Pueblo, and Incan civilizations of Central and South America. The human structure of life in all these civilizations had many similarities: the sense of the divine, ritual forms, social hierarchies, basic technologies, agricultural economies, the temple architecture and sculpture. The great volume of civilizational accomplishments came in this period, especially in the development of the spiritual disciplines whereby human life was shaped in the image of the divine.

But then in the Western world a new capacity for understanding and controlling the dynamics of the earth came into being. While former civilizations established our exalted place within the seasonal sequence of the earth's natural rhythms and established those spiritual centers where the meeting of the divine, the natural, and the human could take place, the new effort, beginning in the sixteenth- and seventeenth-century work of Francis Bacon, Galileo Galilei, and Isaac Newton, was less concerned with such psychic energies than with physical forces at work in the universe and the manner in which we could avail ourselves of these energies to serve our own well-being.

By the mid-eighteenth century the invention of new technologies had begun whereby we could manipulate our environment to our own advantage. At this time also an "objective world" was born—a world clearly distinct from ourselves and available not as a means of divine communion, but as a vast realm of natural resources for exploitation and consumption. These scientific attitudes and technological inventions became the modern substitutes for the mystical vision of divine reality and the sympathetic evocation of natural and spiritual forces by ritual and prayerful invocation.

Yet, though differing in its method, the historical drive of Western society toward a millennium of earthly beatitude remained the same. But the means had changed. Human effort, not divine grace, was the instrument for this paradisal realization. The scientists and inventors, the bankers and commercial magnates, were now the saints who would reign. This, then, was the drive in the technological age. It was an energy revolution not only in terms of the physical energies now available to us, but also in terms of the psychic energies. Never before had we experienced such a turbulent period, such a movement to alter the world, to bring about an earthly redeemed state, and, finally, to attain such power as was formerly attributed only to the natural or to the divine.

This achievement was associated with a sense of political and social transformation that would release us from age-old tyrannies. The very structure of existence was being altered. In this mood America was founded, achieved independence, and then advanced throughout the nineteenth century to a position of world dominance in the early twentieth century. America took seriously the words written by Thomas Paine in 1775 in his pamphlet *Common Sense*: "We have it in our power to begin the world over again. A situation similar to the present hath not appeared since the days of Noah until now. The birthday of a new world is at hand. . . ." Earlier the pilgrims had foreseen this continent as the setting for the new City on a Hill to which the universal human community could look for guidance into a glorious future. Throughout the founding years, even until the twentieth century, peoples crossed the Atlantic with a vivid sense that they were crossing the Red Sea from the slavery of Egypt to the freedom and abundance of the Promised Land.

With such expectation came a new exhilaration in our powers to dominate the natural world. This led to a savage assault upon the earth such as was inconceivable in prior times. The experience of sacred communion with the earth disappeared. Such intimacy was considered a poetic conceit by a people who prided themselves on their realism, their aversion to all form of myth, magic, mysticism, and superstition. Little did these people know that their very realism was as pure a superstition as was ever professed by humans, their devotion to science a new mysticism, their technology a magical way to paradise.

As with all such illusory situations, the awakening can be slow and painful and filled with exaggerated reactions. Our present awakening from this enchantment with technology has been particularly painful. We have altered the earth and human life in many irrevocable ways. Some of these have been creative and helpful. Most have been destructive beyond imagination.

Presently we are entering another historical period, one that might be designated as the ecological age. I use the term *ecological* in its primary meaning as the relation of an organism to its environment, but also as an indication of the interdependence of all the living and nonliving systems of the earth. This vision of a planet integral with itself throughout its spatial extent and its evolutionary sequence is of primary importance if we are to have the psychic power to undergo the psychic and social transformations that are being demanded of us. These transformations require the assistance of the entire planet, not merely the forces available to the human. Otherwise we mistake the order of magnitude in this challenge. It is not simply adaptation to a reduced supply of fuels or to some modification in our system of social or economic controls. Nor is it some slight change in our educational system. What is happening is something of a far greater magnitude. It is a radical change in our mode of consciousness. Our challenge is to create a new language, even a new sense of what it is to be human. It is to transcend not only national limitations, but even our species isolation, to enter into the larger community of living species. This brings about a completely new sense of reality and of value.

What is happening was unthinkable in ages gone by. We now control forces that once controlled us, or more precisely, the earth process that formerly administered the earth directly is now accomplishing this task in and through the human as its conscious agent. Once a creature of earthly providence, we are now extensively in control of this providence.

We now have extensive power over the ultimate destinies of the planet, the power of life and death over many of its life systems. For the first time we can intervene directly in the genetic process. We can dissolve the ozone layer that encircles the earth and let the cosmic radiation bring about distortions in the life process. We can destroy the complex patterns of life in the seas and make the rivers uninhabitable. And we could go on with our description of human power, even over the chemical constitution and the very topography of the planet.

In its order of magnitude, this change in our relation to the earth is much greater than that experienced when the first Geolithic civilization came into being some twelve thousand years ago. Nothing since then, not even the great civilizational structures themselves, produced change on such a significant scale. Such change cannot be managed by partial accommodations or even by major adjustments within the civilizational contexts of the past. The context of survival is radically altered. Our problems can no more be resolved within our former pattern of the human than the problems that led to quantum physics could be dealt with by any adjustment within the context of the Newtonian universe.

No adequate scale of action can be expected until the human community is able to act in some unified way to establish a functional relation with the earth process, which itself does not recognize national boundaries. The sea and air and sky and sunlight and all the living forms of earth establish a single planetary system. The human at the species level needs to fulfill its functional role within this life community, for in the end the human community will flourish or decline as the earth and the community of living species flourishes or declines.

A primary allegiance to this larger community is needed. It will do little good for any nation to seek its own well-being by destroying the very conditions for planetary travel. This larger vision is no longer utopian. It directly concerns the hardest, most absolute reality there is: the reality of the water we drink, the air we breathe, the food we eat.

However ineffective in many of its activities, the United Nations and its constituent members have begun to understand this reality: thus a World Charter for Nature, approved by the Assembly in 1982; thus, also, the megaconferences on environment, technology, habitation, water, and all those basic elements of life that must now attain human protection and just distribution for the welfare not only of the human, but of every living species. Whether as nations or as species, we have a single destiny with the larger community of earthly life.

Much of our trouble during these past two centuries has been caused by our limited, our microphase, modes of thought. We centered ourselves on the individual, on personal aggrandizement, on a competitive way of life, and on the nation, or the community of nations, as the guarantor of freedom to pursue these purposes. A sense of the planet Earth never entered our minds. We paid little attention to the more comprehensive visions of reality. This was for the poets, the romanticists, the religious believers, the moral idealists. Now we begin to recognize that what is good in its microphase reality can be deadly in its macrophase development.

Much of human folly is a consequence of neglecting this single bit of wisdom. A few hundred automobiles with good roads may be a great blessing. Yet when the number increases into the millions and hundreds of millions, the automobile is capable of destroying the higher forms of life on the entire planet. So with all human processes: undisciplined expansion and self-inflation lead only to destruction. Apart from the well-being of earth, no subordinate life system can survive. So it is with economics and politics: Any particular activity must find its place within the larger pattern, or it will die and perhaps bring down the larger life system itself. This change of scale is one of the most significant aspects in the change of consciousness that is needed.

The ecological age into which we are presently moving is an opposed, though complementary, age that succeeds the technological age. In a deeper sense this new age takes us back to certain basic aspects of the universe that were evident to the human mind from its earliest period, but that have been further refined, observed, and scientifically stated in more recent centuries. These governing principles of the universe have controlled the entire evolutionary process from the moment of its explosive origin some fourteen billion years ago to the shaping of the earth, the emergence of life and consciousness, and so through the various ages of human history. These principles, known in past ages by intuitive proc-

esses, are now understood by scientific reasoning, although their implications have not yet been acted upon in any effective way. The ecological
age must now activate these principles in a universal context if the human
venture is to continue. These principles on which the universe functions
are three: differentiation, subjectivity, and communion.

Differentiation is the primordial expression of the universe. In the
fiery violence of some billions of degrees of heat, the original energy
dispersed itself through vast regions of space not as some homogeneous
smudge of jellylike substance, but as radiation and as differentiated particles eventually distributed through a certain sequence of elements, man-
ifesting an amazing variety of qualities. These were further shaped into
galactic systems composed of highly individuated starry oceans of fire.
Everywhere we find this differentiating process taking place. In our own
solar system, within the sequence of planets, we find the planet Earth
taking shape as the most highly differentiated reality we know about in
the entire universe. Life on planet Earth finds expression in an overwhelming variety of manifestations. So, too, with the human: As soon as
we appear, we immediately give to human existence multiple modes of
expression. These themselves change through the centuries.

The second primary principle is that of increased subjectivity. From
the shaping of the hydrogen atom to the formation of the human brain,
interior psychic unity has consistently increased along with a greater com-
plexification of being. This capacity for interiority involves increased unity
of function through ever more complex organic structures. Increase in
subjectivity is associated with increased complexity of a central nervous
system. Then comes the development of a brain. With the nervous system
and the brain comes greater freedom of control over the activity of the
organism. In this manner planet Earth becomes ever more subject to the
free interplay of self-determining forces. With subjectivity is associated the
numinous quality that has traditionally been associated with every reality
of the universe.

A third principle of the universe is the communion of each reality
of the universe with every other reality in the universe. Here our scientific
evidence confirms, with a magnificent overview, the ancient awareness
that we live in a *universe—a single, if multiform, energy event.* The unit
of the entire complex of galactic systems is among the most basic expe-

317

rience of contemporary physics. Although this comprehensive unity of the universe was perceived by primitive peoples, affirmed by the great civilizations, explained in creation myths the world over, outlined by Plato in his *Timaeus,* and given extensive presentation by Newton in his *Principia,* nowhere was the full genetic relatedness of the universe presented with such clarity as by the scientists of the twentieth century.

To Isaac Newton we are especially indebted for our understanding of the gravitational attraction of every physical reality to every other physical reality in the universe, an attraction that finds its ultimate fulfillment in the affective attractions that exist throughout the human community. Without the gravitational attraction experienced throughout the physical world, there would be no emotional attraction of humans to one another. To Darwin we are indebted for our understanding of the genetic unity of the entire web of living beings. To Einstein and his theories of relativity we are indebted for a new awareness of how to think about the dynamics of relatedness in the universe.

The ecological age fosters the deep awareness of the sacred presence within each reality of the universe. There is an awe and reverence due to the stars in the heavens, the sun, and all heavenly bodies, to the seas and the continents; to all living forms of trees and flowers; to the myriad expressions of life in the sea; to the animals of the forests and the birds of the air. To wantonly destroy a living species is to silence forever a divine voice. Our primary need for the various lifeforms of the planet is a psychic, rather than a physical, need. The ecological age seeks to establish and maintain this subjective identity, this authenticity at the heart of every being. If this is so of the prehuman phase of life, it is surely true of the human also.

Only such a comprehensive vision can produce the commitment required to stop the world of exploitation, of manipulation, of violence so intense that it threatens to destroy not only the human city, but also the planet itself. If we would terminate this danger, we must create another, less-vulnerable, life situation. So far, this new world order has not had its adequate presentation. Yet when it comes, it will take the form of what we are designating as the ecological age. This age can provide the historical dynamism associated with the Marxist classless society, the age of plenty envisaged by the capitalist nations, and the millennial age of peace envisaged in the Apocalpyse of John the Divine. At present, however, we are in that phase of transition that must be described as the groping phase. We are like a musician who faintly hears a melody deep within the mind, but not clearly enough to play it through. This is the inner agony we experience, especially when we consider that the music we are creating is the very reality of the universe.

It would be easier if we would remember that the earth itself, as the primary energy, is finding its way both to interior conscious expression in the human and to outer fulfillment in the universe. We can solve nothing by dreaming up some ephemeral structure of reality or by giving the direction of the earth over to our bureaucratic institutions. We must simply respond to the urgencies imposed on us by the energy that holds the stars within the galactic clusters, that shaped the planet under our feet, that has guided life through its bewildering variety of expression, and that has found even higher expression in the exotic tribes and nations, languages, literature, art, music, social forms, religious rituals, and spiritual disciplines over the surface of the planet. There is reason to believe that those mysterious forces that have guided earthly events thus far have not suddenly collapsed under the great volume of human affairs in this late twentieth century.

What is clear is that the earth is mandating that the human community assume a responsibility never assigned to any previous generation. We are involved in a process akin to initiation processes that have been known and practiced from earliest times. The human community is passing from its stage of childhood into its adult stage of life. We must assume adult responsibilities. As the maternal bonds are broken on one level to be reestablished on another, so the human community is being separated from the dominance of Nature on one level to establish a new and more mature relationship. In its prior period the earth acted independently as the complete controlling principle; only limited control over existence was assigned to ourselves. Now the earth insists that we accept greater responsibility, a responsibility commensurate with the greater knowledge communicated to us.

This responsibility has so far been more than we could use wisely, just as the new powers of a young adult are powers seldom used wisely without an intervening period of confusion, embarrassment, and juvenile mistakes. As the child eventually learns a mature mode of conduct, discipline, and responsibility, so now, as individuals and as a planetary community, we will, it is hoped, learn our earth responsibilities.

Our problem is, of course, the problem of recognizing the primacy of the natural world and its spontaneous functioning in all that we do. Our own actions can be truly creative only when they are guided by these deeper spontaneities. Human administration of the universe in any comprehensive manner is far too great a task for any controlled process on the part of humans, just as the movement of the arm to pick up and drink a cup of tea would hardly be possible if we were required to consciously manipulate each movement of the eye, the arm, the nervous system, the oxygen, and the blood flow. Yet we do the act spontaneously, with extensive awareness and control. There is deliberation, but also spontaneity. As with our earth in all its processes, so with the human community: There are inner, spontaneous, all-pervasive forces present that are gradually responding to this integral functioning of the total system. What we need, what we are ultimately groping toward, is the sensitivity required to understand and respond to the psychic energies deep in the very structure of reality itself. Our knowledge and control of the environment is not absolute knowledge or absolute control. It is a cooperative understanding and response to forces that will bring about a proper unfolding of the earth process if we do not ourselves obstruct or distort these forces that seek their proper expression.

I suggest that this is the ultimate lesson in physics, biology, and all the sciences, as it is the ultimate wisdom of tribal peoples and the fundamental teaching of the great civilizations. If this has been obscured by the adolescent aspect of our earliest scientific and technological development, it is now becoming clear to us on an extensive scale. If responded to properly with our knowledge and new competencies, these forces will find their integral expression in the spontaneities of the new ecological age. To assist in bringing this about is the present task of the human community.

List of Contributors

Peter A. A. Berle has been working in environmental and conservation issues for over twenty years as a legislator, litigating lawyer, and administrator. He is currently the president of the National Audubon Society.

Thomas Berry is Director of the Riverdale Center for Religious Research, which is devoted to the study of the classical religious traditions in their relation to the scientific-technological world of the future. He is the author of numerous articles and *The Dream of the Earth* (Sierra Club Books).

Richard S. Booth is a lawyer, an associate professor in the Department of City and Regional Planning at Cornell University, and an environmental activist.

Peter Borrelli is editor of *The Amicus Journal,* a publication of the Natural Resources Defense Council.

Robert H. Boyle was on the staff of *Sports Illustrated* for thirty-two years until his retirement in 1986. He is best known as an environmental watchdog for the Hudson River, and is co-founder and president of the Hudson River Fishermen's Association.

Janet Welsh Brown is Senior Associate at the World Resources Institute.

John N. Cole is a freelance writer and conservationist in Brunswick, Maine. In 1968 and 1969, he was editor of *The Maine Times,* an award-winning weekly in Topsham.

Barry Commoner is a prominent environmental scientist, writer, and lecturer. Currently he is a scientific advisor to the New York State Legislative Commission on Science and Technology and the Committee for Responsible Genetics.

Amos S. Eno is the Director of Policy for the National Fish and Wildlife Foundation. He is the former director of Wildlife Programs for the National Audubon Society and before that was employed in the U.S. Department of the Interior.

Michael Frome is on the faculty of Western Washington University, where he is initiating a program of studies in environmental journalism. The University of Tennessee Press is publishing a collection of his essays, *Conscience of a Conservationist.*

J. William Futrell is president of the Environmental Law Institute, a nonprofit national center for research in environmental law and policy in Washington, DC. He is a past president of the Sierra Club.

Lois Marie Gibbs is founder and executive director of Citizen's Clearinghouse for Hazardous Wastes (CCHW).

Jay D. Hair is the president of the National Wildlife Federation.

Huey D. Johnson is a former Secretary for Natural Resources for the State of California. He is currently the president of The New Renaissance Center in Sausalito.

Charles E. Little is a Washington-based writer on the environment, natural resources, and rural affairs. His most recent book is *Louis Bromfield at Malabar,* published in 1988 by The Johns Hopkins University Press.

Randal O'Toole is the director of Cascade Holistic Economic Consultants (CHEC), a nonprofit forestry consulting firm. He is most recently the author of *Reforming the Forest Service* (published by Island Press, 1988).

Vawter Parker is Coordinating Attorney of the San Francisco–based Sierra Club Legal Defense Fund.

Nathaniel P. Reed is a former Assistant Secretary of the Department of the Interior and a current member of the boards of directors of the National Geographic Society and the Natural Resources Defense Council.

William K. Reilly, Jr. is President of the World Wildlife Fund and The Conservation Foundation.

Dick Russell is a freelance writer and an active conservationist in Boston.

Gus Speth is the President of the World Resources Institute. He is a former chairman of the President's Council on Environmental Quality and was formerly an attorney with the Natural Resources Defense Council.

Karen J. Stults is a Research Associate with Citizen's Clearinghouse for Hazardous Wastes (CCHW).

Fredric P. Sutherland is Executive Director of the San Francisco–based Sierra Club Legal Defense Fund.

Tom Turner is a staff writer for the Sierra Club Legal Defense Fund and a former editor of *Not Man Apart.*

Stewart Udall was Secretary of the Interior from 1961 to 1969, and is the author of *The Quiet Crisis* and *Agenda for Tomorrow.*

Cynthia Wilson is the executive director of Friends of the Earth, an organization committed to the preservation, restoration, and rational management of the earth.

Index

Also Available from Island Press

Land and Resource Planning in the National Forests
By Charles F. Wilkinson and H. Michael Anderson
Foreword by Arnold W. Bolle

This comprehensive, in-depth review and analysis of planning, policy, and law in the National Forest System is the standard reference source on the National Forest Management Act of 1976 (NFMA). This clearly written, non-technical book offers an insightful analysis of the Fifty Year Plans and how to participate in and influence them.

1987. xii, 396 pp., index.
Paper ISBN 0-933280-38-6. **$19.95**

Reforming the Forest Service
By Randal O'Toole

Reforming the Forest Service contributes a completely new view to the current debate on the management of our national forests. O'Toole argues that poor management is an institutional problem; he shows that economic inefficiencies and environmental degradation are the inevitable result of the well-intentioned but poorly designed laws that govern the Forest Service. This book proposes sweeping reforms in the structure of the agency and new budgetary incentives as the best way to improve management.

1988. xii, 256 pp., graphs, tables, notes.
Cloth, ISBN 0-933280-49-1. **$34.95**
Paper, ISBN 0-933280-45-9. **$19.95**

Last Stand of the Red Spruce
By Robert A. Mello
Published in cooperation with Natural Resources Defense Council

Acid rain—the debates rage between those who believe that the cause of the problem is clear and identifiable and those who believe that the evidence is inconclusive. In *Last Stand of the Red Spruce,* Robert A. Mello has written an ecological detective story that unravels this confusion and explains how air pollution is killing our nation's forests. Writing for a lay audience, the author traces the efforts of scientists trying to solve the mystery of the dying red spruce trees on Camels Hump in Vermont. Mello clearly and succinctly presents both sides of an issue on which even the scientific community is split and concludes that the scientific evidence uncovered on Camels Hump elevates the issues of air pollution and acid rain to new levels of national significance.

1987. xx, 156 pp., illus., references, bibliography.
Paper, ISBN 0-933280-37-8. **$14.95**

Western Water Made Simple, by the editors of **High Country News**
Edited by Ed Marston

Winner of the 1986 George Polk Award for environmental reporting, these four special issues of *High Country News* are here available for the first time in book form. Much has been written about the water crisis in the West, yet the issue remains confusing and difficult to understand. *Western Water Made Simple,* by the editors of *High Country News,* lays out in clear language the complex issues of Western water. This survey of the West's three great rivers — the Colorado, the Columbia, and the Missouri — includes material that reaches to the heart of the West — its ways of life, its politics, and its aspirations. *Western Water Made Simple* approaches these three river basins in terms of overarching themes combined with case studies — the Columbia in an age of reform, the Colorado in the midst of a fight for control, and the Missouri in search of its destiny.

1987. 224 pp., maps, photographs, bibliography, index.
Paper, ISBN 0-933280-39-4. **$15.95**

**The Report of the President's Commission on Americans Outdoors:
The Legacy, The Challenge**
With Case Studies
Preface by William K. Reilly

"If there is an example of pulling victory from the jaws of disaster, this report is it. The Commission did more than anyone expected, especially the administration. It gave Americans something serious to think about if we are to begin saving our natural resources."
—Paul C. Pritchard, President,
National Parks and Conservation Association

This report is the first comprehensive attempt to examine the impact of a changing American society and its recreation habits since the work of the Outdoor Recreation Resource Review Commission, chaired by Laurance Rockefeller in 1962. The President's Commission took more than two years to complete its study; the Report contains over sixty recommendations, such as the preservation of a nationwide network of "greenways" for recreational purposes and the establishment of an annual $1 billion trust fund to finance the protection and preservation of our recreational resources. The Island Press edition provides the full text of the report, much of the additional material compiled by the Commission, and twelve selected case studies.

1987. xvi, 426 pp., illus., appendixes, case studies.
Paper, ISBN 0-933280-36-X. **$24.95**

Public Opinion Polling: A Handbook for Public Interest and Citizen Advocacy Groups
By Celinda C. Lake, with Pat Callbeck Harper

"Lake has taken the complex science of polling and written a very usable 'how-to' book. I would recommend this book to both candidates and organizations interested in professional, low-budget, in-house polling."—Stephanie Solien, Executive Director, Women's Campaign Fund.

Public Opinion Polling is the first book to provide practical information on planning, conducting, and analyzing public opinion polls as well as guidelines for interpreting polls conducted by others. It is a book for anyone—candidates, state and local officials, community organizations, church groups, labor organizations, public policy research centers, and coalitions focusing on specific economic issues—interested in measuring public opinion.

1987. x, 166 pp., bibliography, appendix, index.
Paper, ISBN 0-933280-32-7. **$19.95**
Companion software now available.

Green Fields Forever: The Conservation Tillage Revolution in America
By Charles E. Little

"*Green Fields Forever* is a fascinating and lively account of one of the most important technological developments in American agriculture. . . . Be prepared to enjoy an exceptionally well-told tale, full of stubborn inventors, forgotten pioneers, enterprising farmers—and no small amount of controversy."—Ken Cook, World Wildlife Fund and The Conservation Foundation.

Here is the book that will change the way Americans think about agriculture. It is the story of "conservation tillage"—a new way to grow food that, for the first time, works *with*, rather than against, the soil. Farmers who are revolutionizing the course of American agriculture explain here how conservation tillage works. Some environmentalists think there are problems with the methods, however; author Charles E. Little demonstrates that on this issue both sides have a case, and the jury is still out.

1987. 189 pp., illus., appendixes, index, bibliography.
Cloth, ISBN 0-933280-35-1. **$24.95**
Paper, ISBN 0-933280-34-3. **$14.95**

Federal Lands: A Guide to Planning, Management, and State Revenues
By Sally K. Fairfax and Carolyn E. Yale

"An invaluable tool for state land managers. Here, in summary, is everything that one needs to know about federal resource management policies." — Rowena Rogers, President, Colorado State Board of Land Commissioners.

Federal Lands is the first book to introduce and analyze in one accessible volume the diverse programs for developing resources on federal lands. Offshore and onshore oil and gas leasing, coal and geothermal leasing, timber sales, grazing permits, and all other programs that share receipts and revenues with states and localities are considered in the context of their common historical evolution as well as in the specific context of current issues and policy debates.

1987. xx, 252 pp., charts, maps, bibliography, index.
Paper, ISBN 0-933280-33-5. **$24.95**

Hazardous Waste Management: Reducing the Risk
By Benjamin A. Goldman, James A. Hulme, and Cameron Johnson for the Council on Economic Priorities

Hazardous Waste Management: Reducing the Risk is a comprehensive source-book of facts and strategies that provides the analytic tools needed by policy makers, regulating agencies, hazardous waste generators, and host communities to compare facilities on the basis of site, management, and technology. The Council on Economic Priorities' innovative ranking system applies to real-world, site-specific evaluations, establishes a consistent protocol for multiple applications, assesses relative benefits and risks, and evaluates and ranks ten active facilities and eight leading commercial management corporations.

1986. xx, 316 pp., notes, tables, glossary, index.
Cloth, ISBN 0-933280-30-0. **$64.95**
Paper, ISBN 0-933280-31-9. **$34.95**

Án Environmental Agenda for the Future

By Leaders of America's Foremost Environmental Organizations

". . . a substantive book addressing the most serious questions about the future of our resources."—John Chafee, U.S. Senator, Environmental and Public Works Committee. "While I am not in agreement with many of the positions the authors take, I believe this book can be the basis for constructive dialogue with industry representatives seeking solutions to environmental problems."—Louis Fernandez, Chairman of the Board, Monsanto Corporation.

The chief executive officers of ten major environmental and conservation organizations launched a joint venture to examine goals that the environmental movement should pursue now and into the twenty-first century. This book presents policy recommendations for implementing the changes needed to bring about a healthier, safer world. Topics discussed include nuclear issues, human population growth, energy strategies, toxic waste and pollution control, and urban environments.

1985. viii, 155 pp., bibliography.
Paper, ISBN 0-933280-29-7. **$9.95**

Water in the West

By Western Network

Water in the West is an essential reference tool for water managers, public officials, farmers, attorneys, industry officials, and students and professors attempting to understand the competing pressures on our most important natural resource: water. Here is an in-depth analysis of the effects of energy development, Indian rights, and urban growth on other water users.

1985. *Vol. III: Western Water Flows to the Cities*
v, 217 pp., maps, table of cases, documents, bibliography, index.
Paper, ISBN 0-933280-28-9. **$25.00**

These titles are available directly from Island Press, Box 7, Covelo, CA 95428. Please enclose $2.00 shipping and handling for the first book and $1.00 for each additional book. California and Washington, DC residents add 6% sales tax. A catalog of current and forthcoming titles is available free of charge. Prices subject to change without notice.